THE
APOLLO 13
MISSION

THE
APOLLO 13
MISSION

Judy L. Hasday, Ed.M.

Introduction by James Scott Brady,
Trustee, the Center to Prevent Handgun Violence
Vice Chairman, the Brain Injury Foundation

Chelsea House Publishers
Philadelphia

To my best friend Cheryl—thanks for always being there for me.

The author would like to express special thanks to Mr. Eugene Kranz for taking the time to discuss his thoughts about his career with NASA and the Apollo 13 *mission.*

Frontispiece: The Apollo 13 *command module* Odyssey *splashes down at the end of its long and eventful journey through space.*

CHELSEA HOUSE PUBLISHERS

PRODUCTION MANAGER Pamela Loos
ART DIRECTOR Sara Davis
DIRECTOR OF PHOTOGRAPHY Judy L. Hasday
MANAGING EDITOR James D. Gallagher
SENIOR PRODUCTION EDITOR J. Christopher Higgins

Staff for **The *Apollo 13* Mission**
ASSOCIATE ART DIRECTOR Takeshi Takahashi
DESIGNER Keith Trego
PICTURE RESEARCHER Lynn Isaacson Fairbanks

The Chelsea House World Wide Web site address is:
http://www.chelseahouse.com

First Printing

1 3 5 7 9 8 6 4 2

Library of Congress Cataloging-in-Publication Data

Hasday, Judy L., 1957–
The Apollo 13 Mission / Judy Hasday.
  p. cm. — (Overcoming adversity)
Includes bibliographical references and index.
ISBN 0-7910-5310-5 — ISBN 0-7910-5311-3 (pbk.)
1. Apollo 13 (Spacecraft)—Juvenile literature. 2. Project Apollo (U.S.)—Juvenile literature. 3. Space vehicle accidents—Juvenile literature. [1. Apollo 13 (Spacecraft). 2. Project Apollo (U.S.). 3. Space vehicle accidents.] I. Title. II. Series.
TL789.8.U6 A538453 2000
629.45'4—dc21
                                            00-038374

CONTENTS

OVERCOMING ADVERSITY

TIM ALLEN
comedian/performer

MAYA ANGELOU
author

THE *APOLLO 13* MISSION
astronauts

LANCE ARMSTRONG
professional cyclist

DREW BARRYMORE
actress

DREW CAREY
comedian/performer

JIM CARREY
comedian/performer

BILL CLINTON
U.S. president

TOM CRUISE
actor

MICHAEL J. FOX
actor

WHOOPI GOLDBERG
comedian/performer

EKATERINA GORDEEVA
figure skater

SCOTT HAMILTON
figure skater

JEWEL
singer and poet

JAMES EARL JONES
actor

QUINCY JONES
musician and producer

ABRAHAM LINCOLN
U.S. president

WILLIAM PENN
Pennsylvania's founder

JACKIE ROBINSON
baseball legend

ROSEANNE
entertainer

MONICA SELES
tennis star

SAMMY SOSA
baseball star

DAVE THOMAS
entrepreneur

SHANIA TWAIN
entertainer

ROBIN WILLIAMS
performer

BRUCE WILLIS
actor

STEVIE WONDER
entertainer

ON FACING ADVERSITY

James Scott Brady

I GUESS IT'S a long way from a Centralia, Illinois, train yard to the George Washington University Hospital Trauma Unit. My dad was a yardmaster for the old Chicago, Burlington & Quincy Railroad. As a child, I used to get to sit in the engineer's lap and imagine what it was like to drive that train. I guess I always have liked being in the "driver's seat."

Years later, however, my interest turned from driving trains to driving campaigns. In 1979, former Texas governor John Connally hired me as a press secretary in his campaign for the American presidency. We lost the Republican primary to a former Hollywood star named Ronald Reagan. But I managed to jump over to the Reagan campaign. When Reagan was elected in 1980, I was "sitting in the catbird seat," as humorist James Thurber would say—poised to be named presidential press secretary. I held that title throughout the eight years of the Reagan administration. But not without one terrible, extended interruption.

It happened barely two months after the Reagan administration took office. I never even heard the shots. On March 30, 1981, my life went blank in an instant. In an attempt to assassinate President Reagan, John Hinckley Jr. armed himself with a "Saturday night special"—a low-quality, $29 pistol—and shot wildly as our presidential entourage exited a Washington hotel. One of the exploding bullets struck me just above the left eye. It shattered into a couple dozen fragments, some of which penetrated my skull and entered my brain.

The next few months of my life were a nightmare of repeated surgery, broken contact with the outside world, and a variety of medical complications. More than once, I was very close to death.

The next few years were filled with frustrating struggles to function with a paralyzed right side, struggles to speak and communicate.

To people who face and defeat daunting obstacles, "ambition" is not becoming wealthy or famous or winning elections or awards. Words like "ambition" and "achievement" and "success" take on very different meanings. The objective is just to live, to wake up every morning. The goals are not lofty; they are very ordinary.

My own heroes are ordinary folks—but they accomplish extraordinary things because they try. My greatest hero is my wife, Sarah. She's accomplished a lot of things in life, but two stand out. The first has been the way she has cared for me and our son since I was shot. A tremendous tragedy and burden was dropped unexpectedly into her life, totally beyond her control and without justification. She could have given up; instead, she focused her energies on preserving our family and returning our lives to normal as much as possible. Week by week, month by month, year by year, she has not reached for the miraculous, just for the normal. Yet in focusing on the normal, she has helped accomplish the miraculous.

Her other most remarkable accomplishment, to me, has been spearheading the effort to keep guns out of the hands of criminals and children in America. Opponents call her a "gun grabber"; I call her a national hero. And I am not alone.

After a seven-year battle, during which Sarah and I worked tirelessly to educate the public about the need for stronger gun laws, the Brady Bill became law in 1993. It was a victory, achieved in the face of tremendous opposition, that now benefits all Americans. From the time the law took effect through fall 1997, background checks had stopped 173,000 criminals and other high-risk purchasers from buying handguns, and the law has helped to reduce illegal gun trafficking.

Sarah was not pursuing fame, or even recognition. She simply started at one point—when our son, Scott, found a loaded handgun on the seat of a pickup truck and, thinking it was a toy, pointed it at Sarah.

Fortunately, no one was hurt. But seeing a gun nearly bring a second tragedy upon our family, Sarah became determined to do whatever she could to prevent senseless death and injury from guns.

Some people think of Sarah as a powerful political force. To me, she's the person who so many times fed me and helped me dress during my long years of recovery.

Overcoming obstacles is part of life, not just for people who are challenged by disabilities, illnesses, or tragedies, but for all people. No matter what the obstacle—fear, disability, prejudice, grief, or a difficulty that isn't likely to "just go away"—we can all work to make this world a better place.

Apollo 13 *lifts off from Pad 39A at the Kennedy Space Center in Florida at 2:13 P.M. on April 11, 1970.*

1

AN ILL-FATED
MISSION NUMBER

"Do surprises turn you on? Then this is your day for the unexpected."
—horoscope for *Aquarius* in the *Houston Post*, April 13, 1970

VERY EARLY ON APRIL 11, 1970, while most Americans were still asleep, the Kennedy Space Center at Cape Canaveral, Florida, bustled with activity. Located on the northern end of Florida's Atlantic coast, the cape was home to the National Aeronautics and Space Administration (NASA) space program. On this spring Saturday morning a 363-foot-tall Saturn V moon rocket idled on Pad A at Launch Complex 39. Superchilled oxygen and hydrogen fuel emitted vapors into the air from the craft, forming milky white puffs that made it seem as though the rocket were alive and breathing. Except for a flood of powerful spotlights aimed at the rocket, the rest of the launchpad was dark.

Later in the day the Saturn V rocket would carry a three-man crew on a voyage to the moon, the second such trip since the momentous

11

James A. Lovell Jr., one of NASA's most experienced astronauts, had originally been selected to command the three-man crew for the Apollo 14 *mission. Lovell was pleased when his crew's assignment was switched to* Apollo 13, *because it meant he would get his chance to walk on the moon six months earlier.*

first manned lunar landing nine months earlier, in July 1969. That was when the astronauts of *Apollo 11*—Neil Armstrong, Edwin "Buzz" Aldrin Jr., and Michael Collins—achieved the goal President John F. Kennedy had set forth in 1961, when he challenged America to land a man on the moon before the end of the decade.

Kennedy did not live to see Armstrong and Aldrin maneuver their lunar module (LM), named *Eagle*, to an area of the moon known as the Sea of Tranquility. But millions of people around the world watched from their homes as the historic journey was televised. At 4:18 P.M. eastern daylight time (EDT) on July 20, 1969, *Eagle* touched down. Seconds later Armstrong's words were transmitted back to Earth: "Houston, Tranquility Base here. The *Eagle* has landed." Countless hours of design, engineering, building, and testing by NASA and its contractors had been successful. The United States had put men on the moon, more than 200,000 miles from Earth.

As dawn spread across Cape Canaveral on April 11, 1970, the astronauts of the *Apollo 13* mission were already preparing themselves for liftoff. In their comfortable quarters at the Kennedy Space Center, Donald "Deke" Slayton, a former astronaut who had been promoted to director of flight crew operations, roused the men from sleep. *Apollo 13* mission commander James A. Lovell Jr., a veteran of three previous space missions, quickly rose to get ready for the launch; Lovell's crewmates, LM pilot Fred W. Haise Jr. and command module (CM) pilot John L. "Jack" Swigert Jr., followed.

The men took the last hot showers they would have for 10 days, and then visited the mission's medical team for a final checkup. They joined Slayton and the *Apollo 13* backup crew for the traditional send-off breakfast of steak and eggs, and then began preparing for the launch.

Soon after, Lovell, Haise, and Swigert emerged from the dressing area of the Space Center in their finely tuned but bulky space suits and helmets. They smiled, waved, and gave cheerful thumbs-up signs to the crowd gathered outside the building before they were whisked away in an air-conditioned van to Pad 39A. There an assistance crew accompanied the astronauts on a 33-story gantry elevator ride to the space capsule. After the three had climbed into the capsule cockpit, the crew closed and sealed the hatch. *Apollo 13* was less than three hours from takeoff.

So far the mission was proceeding smoothly, despite the mutterings of superstitious types who believed that the mission number—13—made it a doomed flight. Some pointed out that the entire mission had ominous numerological overtones. When the digits contained in the date of the scheduled launch—4/11/70—were added together, for example, they totaled 13. Furthermore, *Apollo 13* was scheduled to lift off at 1:13 P.M. Houston time—1313 by military clocks. (The launch took place in Florida at 2:13 P.M. eastern standard time [EST], one hour later than in Houston, where Mission Control was located.) And if the mission continued on schedule, *Apollo 13* would enter the moon's gravitational field on April 13.

Of course, there had been a few problems with the mission, but that was not unusual. For example, the mission's original CM pilot, Thomas "Ken" Mattingly II, with whom Lovell and Haise had trained for two years, was grounded seven days before the launch after doctors discovered that he'd been exposed to measles. Then, just before launch, ground tests indicated a possible problem with the insulation in a helium tank on the LM. Still, these were not earth-shattering problems, and NASA officials,

The Apollo 13 *astronauts selected this emblem for their mission. Based on Roman mythology, it depicts horses pulling the chariot of the Sun God (Apollo) from the earth to the moon. The Latin motto means, "From the moon, knowledge."*

and even Lovell himself, were more amused than worried by the supposed omens surrounding the mission. They were not about to let such beliefs affect hard, quantifiable, scientific logic. In the book *Lost Moon*, Jim Lovell describes the prevailing sentiment: "Bringing one of humanity's greatest scientific endeavors eyeball to eyeball with one of its most enduring superstitions had an irresistible appeal, and most people applauded the hubris [exaggerated pride], the c'mon-I-dare-you arrogance, of flying the mission anyway, and even embroidering a big loud 'XIII' on the patches of the suits the astronauts would be wearing throughout the flight."

The countdown to launch proceeded without incident as Lovell, Haise, and Swigert completed their system checks. This mission did not attract the huge crowds that had crammed every inch of the cape and its surrounding area when *Apollo 11* had taken off for the moon. The nation's interest in the space program had begun to wane somewhat—after all, how do you continue to generate enthusiasm once the goal of landing on the moon has been reached?

The Apollo program may have begun to lose its attraction for the public, but for Jim Lovell it was a dream come true. *Apollo 13* would probably be his last mission, and commanding the flight that would take him to the moon— and actually walking on the moon, as only four other human beings had to that point—was an unimaginable thrill. For Haise and Swigert there was nothing humdrum about this mission either. It was their first and they had trained long and hard to get where they were. Now it was actually about to happen. They were going to the moon!

With all systems "go" (ready for launch), the final countdown began. At T-minus-8.9 seconds (8.9 seconds before launch), *Apollo 13*'s ignition sequence began. At T-minus-2 seconds all five S-IC engines ignited. The Saturn V, with the crew of *Apollo 13* atop it, began to lift off amid a thunderous, roaring crescendo and a huge cloud of smoke and flames, clearing the tower and soaring skyward, away from Earth.

Five minutes into the flight the astronauts felt a slight vibration, and then one of the five S-II booster engines cut off two minutes earlier than scheduled. It wasn't a problem, though: the four remaining engines burned 34 seconds longer to compensate, and the booster, with plenty of fuel, burned an additional nine seconds to put the craft into orbit. Three hours later the crew of *Apollo 13* blasted out of Earth's orbit and headed toward the moon.

The following morning, on the first full day in the cozy capsule, which was heated to a balmy 72°F, a relaxed Lovell contacted the ground crew's Capsule Communicator (Capcom), Joe Kerwin, to ask about the day's news. Kerwin began scanning the Sunday paper for headlines, and he read them to the astronauts, perched in their capsule thousands of miles above Earth:

The [Houston] Astros survived, 8 to 7. . . . They had earth-quakes in Manila and other areas of the island of Luzon. West German Chancellor Willy Brandt, who witnessed your launch from the Cape yesterday, and President Nixon will complete a round of talks. The air traffic controllers are still out [on strike], but you'll be happy to know the controllers in Mission Control are still on the job.

Kerwin's comment cracked up the Apollo crew. Then, as if setting up another punch line, Kerwin mentioned that most Americans were spending the day completing their income tax return forms, which are due annually on April 15. Kerwin then paused for effect before saying, "Uh-oh, have you guys completed your income tax?" Swigert immediately cut in, wanting to know how to file for an extension. In the rush to prepare for his last-minute replacement of Ken Mattingly, Swigert had forgotten to file his tax return. The flight crew could hear members of the ground crew chuckling. Swigert was not amused; he really hadn't had time to file. Finally, Kerwin told Swigert he would see what they could do about his problem.

Most of day three of the *Apollo 13* mission—Monday, April 13—passed uneventfully. That evening the astro-nauts were scheduled to broadcast live images from the CM. Amazingly, none of the networks believed that the live broadcast from space was worth a disruption in their regular programming. The space telecast, scheduled to begin at 8:24 P.M. EST, would cut into NBC's *Rowan & Martin's Laugh-In*, CBS's *Here's Lucy*, and ABC's airing of the 1966 movie *Where the Bullets Fly*. Even though the broadcast from *Apollo 13* would not be aired on the net-works, NASA decided to proceed with it anyway, assum-ing that TV stations would want to air taped highlights of the telecast during their nightly newscasts.

Besides the NASA crew members working their shifts, several of the astronaut's family members were on hand in the space center's VIP gallery to see the 45-minute pro-gram: Lovell's wife Marilyn, his daughters Barbara and

Susan, and Haise's wife Mary were about to see their husbands' and father's images beamed down from space for the first time.

The opening image was that of a fuzzy but relaxed-looking Fred Haise, floating in the zero gravity of the tunnel between the CM (named *Odyssey*) and the LM (named *Aquarius*). Then Lovell, who was operating the camera, began to give viewers a tour of *Odyssey* and *Aquarius*. The mission commander and Haise explained the equipment aboard *Aquarius* that would be used for the trip to and from the lunar surface.

While Lovell's camera beamed pictures back to Earth, Mission Control planned chores and maneuvers that the astronauts would have to execute before they signed off for the evening. Seymour "Sy" Liebergot, the electrical and environmental command officer (EECOM) at Mission

Technicians in the Mission Operations Control Room in Houston, Texas, keep busy during the April 13 television transmission from Apollo 13. On the screen to the right is astronaut Fred Haise. The television networks opted not to interrupt their regular programming to broadcast the signals from space.

Control, requested a "cryo stir" of all four fuel tanks (two oxygen, two hydrogen) at 55 hours, 50 minutes into the mission. This was a routine exercise. The oxygen and hydrogen were maintained in a superfrigid, cryogenic state, which is not quite a solid, a liquid, or a gas. The elements, kept in separate tanks, were channeled through three fuel cells connected to the tanks and hooked up to electrodes. When the oxygen and hydrogen flowed into the fuel cells and reacted with the electrodes, they generated a chemical reaction that produced electricity, water, and oxygen—the fundamentals for sustaining life.

Since the oxygen used by the fuel cells and the oxygen used to maintain life support came from the same cryogenic tanks, it was an especially precious commodity. Inside each tank were two electrical probes. One was designed to gauge the temperature and pressure in the tanks; the other was a combination heater and fan. The heater was used to expand the oxygen if the tank pressure dropped too low, and the fan was used to "stir up" the oxygen to keep it in the cryogenic state. The cryo stir was a chore requested each day by the EECOM on shift.

After Lovell had been broadcasting for more than 25 minutes, Mission Control suggested that he wrap up his program. As Haise finished up his own chores on the spacecraft, he pushed the repress valve, which maintained the air pressure between *Odyssey* and *Aquarius*. The valve let out a hiss and a thud, familiar sounds to Lovell and Swigert but jolting nonetheless. It seems that Haise had become somewhat of a practical joker and would sometimes press the valve just to startle his crewmates. "Every time he [did] that," Lovell explained in *Lost Moon*, "our hearts jump[ed] into our mouths."

After concluding the TV broadcast, Lovell began preparing maneuvering instructions from Capcom Jack Lousma. Haise had finished closing up the LM and was on his way through the tunnel connecting it to *Odyssey*. Following EECOM Liebergot's request to perform the cryo stir, Swigert flipped the switch that turned on the fans in the four tanks.

Sixteen seconds later the astronauts heard a loud boom and felt the ship shudder. Haise saw the walls in the tunnel shake, and he heard what sounded like metal twisting from side to side. Both Swigert and Lovell felt the craft quake beneath them. At first Lovell thought it must have been Haise playing another joke with the repress valve. But the LM pilot's eyes were "wide as saucers," and Lovell realized that the man had no idea what had just happened.

Haise mumbled, "It wasn't me," and Lovell's gaze shifted to Swigert's face, where he saw the same look of dismay. It was obvious that the event had shaken both men. Swigert looked at the extensive instrument panels above his head and saw an amber light flashing. Just then, a warning signal went off in Haise's headphones. Now truly alarmed, Swigert continued scanning the cockpit controls for some type of explanation and noticed that the craft had suffered a severe and sudden drop in power. The light on the main B bus panel—one of two main power distribution panels, each of which provided 50 percent of the CM's power—began flashing. Something was terribly wrong.

The *Apollo 13* mission, a launch that had begun routinely two days earlier as "just another moon mission," was about to capture the attention of the entire world.

A close-up view of the far side of the moon, taken during one of the Apollo missions. For centuries, man has wondered about Earth's satellite and dreamed of one day walking on its surface.

2

"WE CHOOSE TO GO TO THE MOON"

"There are things that are known and things that are unknown; in between is exploration."

—author unknown

FOR CENTURIES MAN has gazed up at the night sky in curiosity and wonder. Ancient astronomers observed that some points of light appeared to move over the course of days and months. They called the moving objects planets (from the Greek word for "wanderers") and named them after Roman gods like Jupiter (king of the gods), Mars (the god of war), Venus (the god of love), Saturn (the god of agriculture), and Mercury (the god of commerce and travel and the messenger to other gods). Stargazers also noticed that some of these lights shot across the sky, and they called them comets. Other orbs that seemed to fall out of the sky were called meteors.

Perhaps the most fascinating light in the sky has been the one closest to Earth, the one most visible to the naked eye: the moon. From

Earth it is second only to the sun itself in brightness, and it seems to have enormous powers. It has inspired lovers, moved oceans, and awed mankind with the beauty of its phases and eclipses. An ancient story tells of Selene, the Greek goddess of the moon, who visited Endymion, her greatest love, each night while he slept and "kissed" him with rays of moonlight. For generations children have delighted in finding a human face, a "man in the moon," in the craggy lunar surface.

Tall tales have been written about journeys to other worlds and about the vehicles needed to get there. In the second century Lucian of Samosata wrote one of the earliest stories about traveling in space, *Vera Historia* (*True History*). In it a sailing vessel and its crew, caught in a storm at sea, are lifted by powerful winds and deposited on the moon. There they encounter the Hippogypi, lunar inhabitants who ride on the backs of three-headed vultures.

Probably the best-known early fictional account of space travel is *From the Earth to the Moon*, written by Jules Verne in 1865. In a story that includes startling similarities to the actual Apollo space program established nearly a century later, Verne creates three astronauts in Florida who are loaded into a small, cylindrical craft called the *Columbiad*, lodged inside the barrel of a huge cannon aimed at the sky. The story ends with the cannon being fired, launching the craft into space. In a sequel, *Round the Moon*, Verne continues the story of the *Columbiad*'s journey. The spacecraft does not make a lunar landing, but it does orbit the moon before heading back to Earth and splashing down in the ocean. Less than 100 years after Verne's stories were published, three American astronauts boarded the *Apollo 8* capsule and were launched into space by a huge rocket. After orbiting the moon, they splashed down in the Pacific Ocean— much like Verne's fictional characters had done.

The idea of space travel eventually moved off of the pages of novels and into the notebooks of early scientists

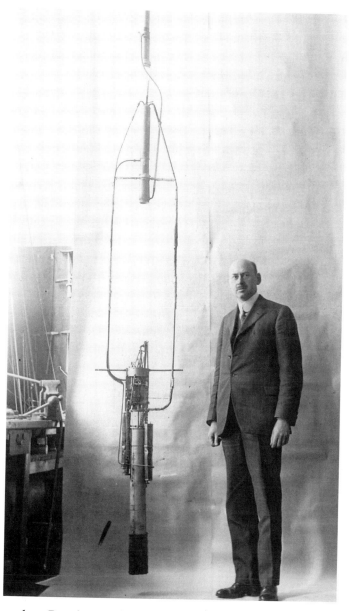

Dr. Robert H. Goddard, with a prototype liquid-propellant rocket. Known as the "Father of American Rocketry," Dr. Goddard's research helped to bring about the Space Age.

such as Russian teacher and self-taught physicist Konstantin Tsiolkovsky (1857–1935), American professor and physicist Robert Goddard (1882–1945), and German physicist Hermann Oberth (1894–1989), all pioneers of the space age.

Tsiolkovsky, an innovative writer, was fascinated with the concept of space exploration. In his book *Exploration of Cosmic Space with Reactive Devices*, Tsiolkovsky not only wrote about rockets, but also described the many other elements needed for interplanetary exploration, such as space suits, lunar stations, and space labs.

Goddard, known as the father of American rocketry, was also inspired by the science fiction stories he'd read as a child. Preoccupied with thoughts of space travel as a teen, he successfully launched the first liquid-propellant rocket from his aunt Effie's backyard on March 16, 1926. Though the rocket remained in the air for only 2.5 seconds, it reached an altitude of 41 feet—the first 41 feet toward space. Over the next 15 years Goddard worked to develop the foundation of the rocketry technology that is used to this day.

Oberth, a physics professor like Goddard, read Tsiolkovsky's writings and studied Goddard's liquid-fuel rocket research. He published his own work, *The Rocket into Interplanetary Space*, in 1923. In the book he combined the concepts of space travel and liquid-fueled rocketry. He was instrumental in bringing the concept of space travel to the attention of scientists and engineers around the world, and he helped found Germany's Verein für Raumschiffahrt (VfR), the Society for Space Travel. Oberth also worked as a technical adviser on a German film called *Frau im Mond* (*Lady in the Moon*); like Jules Verne's written tales, the movie generated widespread interest in space travel.

A young German fellow named Wernher von Braun read Oberth's book two years after it was published. Wernher's mother had given him a telescope so he could study the night sky, and he read many articles that described explorations to the moon and beyond. The thought of such expeditions thrilled the young man: "Interplanetary travel! Here was a task worth dedicating one's life to! Not just to stare through a telescope at the moon and the planets but to soar through the heavens and actually explore the mysterious universe!"

One story tells how 13-year-old von Braun strapped six skyrockets to a toy wagon, lit the fuses, and watched as the wagon roared off. The makeshift "rocket" traveled five blocks before it exploded. All that remained when the smoke cleared was a charred lump. Not to be deterred, von Braun learned calculus and trigonometry, then studied physics before he joined Oberth's VfR, which by 1930 was among the most important rocket experiment societies in Europe or the United States. On any given day one could hear the roar of a VfR test rocket coming from the *Raketenflugplatz* (rocket flying field), a wooded area on the outskirts of Berlin.

Just before World War II, German authorities, interested in rocketry for military purposes, offered von Braun and his colleagues the opportunity to work in government research laboratories. Von Braun was appointed head of the military rocket program. For the next 13 years he and his team perfected the A-4 military rocket, later renamed V-2, the *V* standing for its main purpose—vengeance. On October 3, 1942, the first 46-foot V-2 lifted off and reached an altitude of 50 miles.

That evening at dinner German military commander General Walter Dornberger toasted von Braun and his brilliant team of technicians and engineers. "Today, the spaceship was born!" Dornberger declared. Von Braun is said to have wryly replied, "If so, it landed on the wrong planet." Over a six-month period, from the fall of 1943 to the spring of 1944, more than 100 perfected V-2s, each armed with one ton of explosives, were launched at London by Adolph Hitler's forces.

Despite Germany's new weaponry, Hitler was losing the war. In March 1945 von Braun and many of his colleagues surrendered to the American forces and their allies, which were closing in on Germany from the west. Other scientists chose to defect to the Soviet Union and work with its

An American version of the V-2 rocket, developed for Nazi Germany by Wernher von Braun, is launched at White Sands Missile Range in New Mexico. After von Braun and other German rocket scientists surrendered at the end of the war, they were brought to the United States to help develop an American rocket program.

growing rocket program. Years later, von Braun explained that he wanted to go to America not only because it "had a reputation for having an especially intense devotion to individual freedom and human rights," but also because in the future he wanted to be on the "winning side."

Von Braun and 127 members of his former team went to work for the U.S. military and continued perfecting rockets at White Sands Proving Ground, a military installation in the New Mexico desert. In 1950 von Braun moved his team to Huntsville, Alabama, to work for the army's Redstone Arsenal Guided Missile Division; later he became director of development at the Ballistic Missile Agency in Huntsville. The Redstone rocket would ultimately allow America to begin its journey into space.

While von Braun was setting the United States on a course with the skies, his Russian counterpart, Sergei Korolev, was doing the same in the Soviet Union's rocket program. The Cold War, the struggle between the United States and the Soviet Union for political and military dominance, was the driving force behind the Soviet's push to develop long-range missiles capable of carrying armed nuclear warheads.

The U.S. government's interests were similar, although it also aimed to pursue scientific information about the earth and its atmosphere. Although America announced plans to launch the first artificial satellite into orbit by the end of 1958, the Soviet Union beat them to this goal. On October 4, 1957, the Soviets stunned the world when the newspaper *Pravda* headlined a successful launch of the world's first artificial satellite. *Sputnik 1* (*sputnik* is a Russian word for traveler), *Pravda* announced, was already in orbit around the earth.

A month later the Soviets achieved another first with a rocket six times heavier than *Sputnik 1*. The launch of *Sputnik 2* was extraordinary for another reason as well: it carried the first living space traveler, a dog named Laika. Caught unaware, and now lagging behind in the "space race," America scrambled to catch up with the Soviet Union. To achieve this, it turned to a well-established military agency—one that had been around for nearly half a century.

The government-sponsored National Advisory Committee for Aeronautics (NACA), formed in 1915, was instituted to conduct research in aerodynamics and to ensure the continual development of air technology—a crucial component of any nation's defense system. Among the distinguished Americans who had served on the committee were airplane pioneer Orville Wright, legendary aviator Charles Lindbergh, and World War I ace fighter pilot Eddie Rickenbacker. NACA's research took place at the Langley Memorial Laboratory in Hampton, Virginia, a small facility that eventually grew into one of the world's largest aviation research centers. For many at NACA, flight experimentation in the sky was simply an extension of similar testing on the ground. It was on one of NACA's experimental flying fields, for example, that test pilot Chuck Yeager took off in the Bell X-1 jet and became the first human being to fly faster than the speed of sound.

NACA officials were certainly aware that space flight would become an outgrowth of aviation development, but they believed it would have to be a gradual process, developed over years of in-depth research, design, and testing. But that strategy changed drastically after the successful launches of *Sputnik 1* and *Sputnik 2*. On July 29, 1958, President Dwight Eisenhower signed Public Law 85-568, creating a new space agency called the National Aeronautics and Space Administration—NASA. The National Advisory Committee for Aeronautics was absorbed by the new agency, which was established as a civilian government space program that would be unconnected to military endeavors. NASA's job was to pursue the exploration of space and to obtain scientific knowledge with peaceful intentions.

To be sure, the new agency worked far differently from NACA's steady but plodding pace. NASA exploded into being, with 8,000 men and women reporting to work on October 1, 1958. Finally things seemed to be looking up— and going up—for the United States.

By that time, America's first satellite had already been launched. After a few dismal failures on the launchpad, von Braun and his team had successfully sent up *Explorer 1* on January 31, 1958, from Cape Canaveral, Florida. Atop a Juno I rocket, *Explorer 1* sailed into Earth's orbit.

But *Explorer 1*'s mission turned out to be far more than

an achievement of orbit. Instruments on board the vehicle had discovered a band of charged radiation particles—the Van Allen belt—encircling the earth. The Soviets might have sent up the first satellite, but America had made the first scientific discovery of the space age.

NASA's next goal was to complete Project Mercury, the first U.S. manned space program. Initially, some officials believed the ideal candidates for the program should be chosen from among daredevils—athletes, race car drivers, and others who were accustomed to taking risks. But President Eisenhower disagreed. He argued that stable, college-educated men (no women were considered) were preferable, since they would not be security headaches. Ultimately, NASA decided to select its astronauts from a pool of military test pilots.

The candidates underwent several screenings, including height, weight, and age restrictions. Of the 508 initial recruits, 32 men were selected. They endured a series of rigorous physical tests and a battery of psychological evaluations. From this group, seven men were chosen.

On April 9, 1959, NASA introduced the seven Mercury astronauts to the press and the American public: air force captains Donald "Deke" Slayton, Virgil I. "Gus" Grissom, and L. Gordon Cooper Jr.; John Glenn Jr., a lieutenant colonel in the marines; navy lieutenant M. Scott Carpenter; and navy lieutenant commanders Walter M. Schirra Jr. and Alan Shepard. All were either seasoned combat aviators or experienced test pilots. Now that the Mercury astronauts had been selected and the first space capsule was in design, the United States could enter the space race wholeheartedly.

On April 12, 1961, however, the Soviets won out again. Red Air Force cosmonaut Yury Gagarin, 27, piloted the *Vostok 1* rocket into orbit around the world, then parachuted down to Earth, landing in a field near the town of Saratov. The news of Gagarin's successful orbit was met with great disappointment at NASA. The United States was

The seven astronauts selected for Project Mercury, the United States's first foray into space, included (left to right) M. Scott Carpenter, L. Gordon Cooper Jr., John H. Glenn Jr., Virgil I. "Gus" Grissom, Walter M. Schirra Jr., Alan B. Shepard Jr., and Donald K. "Deke" Slayton.

clearly behind in space technology. Nonetheless, NASA doggedly continued preparing for its first Mercury mission.

Shortly after sunrise on May 5, 1961, astronaut Alan Shepard climbed aboard the *Freedom 7* capsule (numbered after the seven Mercury astronauts) and awaited ignition countdown. Despite several delays, Shepard blasted off at 9:34 A.M. The flight was brief—only 15 minutes long—but America had at last put a man into space. Shepard's flight was suborbital, meaning that *Freedom 7* flew up through the atmosphere into space, then returned to earth without orbiting the planet.

A few weeks after Shepard's historic ride, President Kennedy convened a special joint session of Congress to address what he termed "urgent national needs." Before he concluded, he turned his attention to the space race. "Space is open to us now," he declared, "and our eagerness to share its meaning is not governed by the efforts of others. We go into space because whatever mankind must undertake, free men must fully share." Kennedy then presented a challenge: "I believe that this nation should commit itself to achieving the goal, before this decade is out, of landing a man on the moon and returning him safely to Earth."

Kennedy's goal was ambitious, but NASA refused to be overcome by pressure. Like a child learning to walk, it took the space race one step at a time. Astronaut Gus Grissom was tapped to man the second suborbital flight of the Mercury project. His flight on *Liberty Bell 7* went well, but a technical malfunction caused the escape hatch to blow prematurely during recovery efforts. Though the capsule sank to the ocean floor (it was recovered in early 2000), Grissom was plucked from the water and rescued.

Following the flight of the *Liberty Bell*, NASA named John Glenn as the first American astronaut to orbit the Earth. On February 20, 1962, aboard *Friendship 7*, Glenn blasted off and achieved orbit. He circled the earth three times before splashing down in the Atlantic Ocean.

In all, six of the seven original Mercury flights were completed. Like Shepard, Grissom, and Glenn, astronauts Carpenter, Schirra, and Cooper all had their turns flying in space. Only Deke Slayton, who was grounded after doctors detected a periodic irregular heart rhythm, never made it into space as a Mercury astronaut.

In a speech at Rice University in September 1962, President Kennedy again spoke of going to the moon. "Why choose this as our goal?" Kennedy asked rhetorically. Then he answered his own question:

During his May 25, 1961, message to a joint session of Congress, President John F. Kennedy declared, "I believe this nation should commit itself to achieving the goal, before this decade is out, of landing a man on the moon and returning him safely to Earth."

We choose to go to the moon in this decade and do the other things not because they are easy, but because they are hard, because that goal will serve to organize and measure the best of our energies and skills, because that challenge is one that we are willing to accept, one we are unwilling to postpone, and one which we intend to win.

Kennedy closed his speech by reminding the audience of British mountain climber George Mallory's reply to those who asked why he wanted to climb Mount Everest.

"Because it is there," Mallory said simply. Kennedy continued, "Well, space is there, and we're going to climb it, and the moon and the planets are there, and new hopes for knowledge and peace are there."

Project Mercury was a success, but several more years would pass before an American would set foot on the moon. NASA's space program became a three-part mission: after the Mercury project, the Gemini program would begin. Its primary goals were to test new equipment and gain more hands-on experience in maneuvers such as spacewalking and positioning the craft for orbital rendezvous and docking. There was much work to do before the third and final part of the space program, Project Apollo, could begin. And NASA had only seven years to complete it.

The first images of Earthrise, showing the luminous Earth rising over the moon, were beamed to Earth by the crew of Apollo 8 *on Christmas Eve 1968. The mission's crew, Frank Borman, Jim Lovell, and Bill Anders, became the first men to orbit the moon, more than 200,000 miles away from Earth.*

3

"ONE SMALL STEP FOR MAN, ONE GIANT LEAP FOR MANKIND"

"I had the ambition to not only go farther than man had gone before, but to go as far as it was possible to go."
—18th-century English navigator Captain James Cook

WHEN NASA INITIATED Project Mercury in 1958, the goals of the man-in-space program were threefold: to successfully orbit a manned spacecraft around the Earth; to study man's ability to function in space and determine the physiological effects on the human body; and to devise a plan to recover both astronauts and space capsules safely. The last of the seven scheduled manned space flights was completed May 15–16, 1963, when L. Gordon Cooper achieved 22 orbits during a 34.5-hour flight aboard *Faith 7*. The success of the Mercury project confirmed that it was possible for humans to survive not only the weightless environment of space, but also the physical stresses of launch and reentry.

The route that took mankind to the moon may have begun with

Project Mercury, but Project Apollo—a series of collaborative efforts by astronauts and hundreds of NASA scientists, engineers, and outside contractors—would take it to its destination. Even before the Mercury missions were under way, NASA engineers organized into the Space Task Group had begun planning a manned moon landing.

The first step in completing such an ambitious project was to determine exactly how a lunar landing would take place. Three proposals were actively considered: the direct ascent mode, the Earth orbit rendezvous, and the lunar orbit rendezvous. Of these, the latter, called the LOR, initially encountered the most resistance from officials. The brainchild of NASA engineer John Houbolt, the LOR required several complex flight maneuvers and two space vehicles, one of which would remain in orbit around the moon, while the other, a much lighter craft, transported two astronauts from lunar orbit to the moon's surface. After both men were safely back on board the moon-landing vehicle, Houbolt proposed, the craft would blast off from the moon and reconnect with the orbiting vehicle. Then the astronauts would climb into a capsule attached to the orbiting vehicle to return to Earth.

Despite initial skepticism over Houbolt's plan, NASA officials ultimately determined that fuel conservation and weight concerns made LOR the best choice. Houbolt's plan, perhaps one of the most overlooked contributions to the entire space program, became the standard for moon landings from the first mission to the moon.

But after NASA settled on LOR, hundreds of questions remained. Could a rendezvous with another orbiting craft even be achieved in space at speeds of 17,000 miles per hour? Could human beings endure a two-week space journey (the amount of time needed to reach the moon and return)? Could astronauts leave the safe, balanced environment of their capsule and step into the void of space to conduct extravehicular activity (EVA, commonly called a space walk) protected only by a relatively thin space suit?

Since there was literally no way on Earth to determine the answers to these questions, NASA initiated an extended series of test journeys in space. The link that joined the pioneering efforts of Project Mercury with the ambitious goals of Project Apollo was aptly called Project Gemini—the name of the third constellation of the zodiac, which features the twin stars Castor and Pollux.

Project Gemini was fashioned as a series of 12 two-man missions, each of which built upon the original goals of Project Mercury: exposing astronauts to weightlessness for longer periods, mastering rendezvous and docking maneuvers with orbiting vehicles, and improving methods of reentry to allow for preselected landing locations. While Mercury was in progress, NASA continued recruiting astronauts, and by September 1962 it announced that nine more astronauts had joined the program: Neil Armstrong, Frank Borman, Charles "Pete" Conrad, Jim Lovell, James McDivitt, Elliot See, Thomas Stafford, Edward White II, and John Young. The following year 14 more men joined the program, including Buzz Aldrin, William Anders, Roger Chaffee, and Michael Collins. And by 1966 NASA had selected its final group of pilot astronauts for the Apollo program. Among the 19 newest members were Fred Haise, Jack Lousma, Ken Mattingly, and Jack Swigert. Some of these men, along with the original Mercury Seven, would first train for missions in the Gemini program.

The Gemini project was a spectacular success in every regard. It gave NASA the answers they needed to safely proceed with the Apollo program. Some of Project Gemini's high points included mission number four—in which Ed White achieved the first walk in space by an American—and the first rendezvous mission, realized by Walter Schirra and Thomas Stafford in *Gemini 6* and Frank Borman and Jim Lovell in *Gemini 7*. Borman and Lovell's record-breaking two weeks in space put to rest concerns about extended space travel.

Later missions helped NASA work out problems in

During Project Gemini the American astronauts proved it was possible for two spacecraft, traveling at more than 17,000 miles per hour, to rendezvous. Gemini 6, *with Walter Schirra and Thomas Stafford aboard, and* Gemini 7, *containing Frank Borman and Jim Lovell, were steered to within three feet of each other. This photo was taken from* Gemini 6.

rendezvous and docking procedures with an additional vehicle in space. The astronauts on these missions practiced their techniques with an unmanned Agenda-D rocket. In the latter half of 1966, NASA sent up four Gemini flights; in the last of these, *Gemini 12*, Buzz Aldrin completed a flawless EVA with Jim Lovell at the helm. The two capped a very successful interim step in man's quest to reach the moon. Thrilled with the success of the Gemini project, NASA was excitedly optimistic about Project Apollo. The possibilities of the moon program seemed almost limitless. But before long, those hopes would be dashed in a fiery tragedy.

Just 10 weeks after the last Gemini flight was completed, the crew of the first Apollo mission began preparing for the maiden flight of the new capsule and rocket designed specifically to take astronauts to the moon and back. The command pilot for *Apollo 1* was Gus Grissom, a veteran of both Mercury and Gemini. Joining Grissom were *Gemini 4* spacewalker Ed White and rookie pilot Roger Chaffee.

But Spacecraft 012 (the production number of the craft) was plagued by problems almost from the start. Many of the capsule's malfunctions were traced to the environmental control system (ECS), which regulated temperature, pressure, and atmosphere inside the CM. There were also countless glitches in the communications systems. As problems mounted, Grissom grew weary and irritable with the mission's progress. Spacecraft 012, he believed, was an accident waiting to happen—and it did happen, with catastrophic consequences.

On the afternoon of January 27, 1967, Grissom, White, and Chaffee were sealed inside the capsule, sitting atop the rocket, to execute a practice countdown (a kind of dress rehearsal before the actual launch, which was scheduled for the following month). At 6:31 P.M., more than five hours after the crew had entered the capsule, they were still struggling with communications problems. Suddenly, Grissom yelled into the radio: "We've got a fire in the cockpit!"

The three astronauts vainly tried to open the escape hatch, but the capsule was already engulfed in flames. Intense heat and smoke prevented rescue teams from getting to the capsule immediately. Five crucial minutes ticked away before they could get the hatch open.

It was too late. Inside, the lifeless bodies of Gus Grissom, Ed White, and Roger Chaffee lay where the men had been seated, amid melted, twisted metal and charred black walls. The three men had become the first fatalities of the NASA space program.

In the aftermath of the deadly fire, NASA administrator James Webb immediately established an accident investigation board to determine its cause and make recommendations to correct the problems. On April 10, 1967, seven days after George Low replaced Joe Shea as head of the Apollo spacecraft program, the investigation board turned in its official report on the *Apollo 1* fire. The source of the fire: an electric spark that had originated in faulty wiring had ignited the oxygen-pure air inside the capsule. But during its painstaking investigation, the board also discovered scores of design flaws and incidents of sloppy workmanship, particularly in the capsule's electrical wiring. As a result of the findings, NASA implemented more than 100 design changes and safety procedures. Among the most important was the installation of a new type of hatch that would open outward faster (the complicated hatch in *Apollo 1* had prevented the astronauts from exiting quickly after fire broke out). Also added were better-insulated wiring and fire-resistant capsule materials, and NASA did away with practice countdowns conducted while astronauts were sealed in capsules filled with pure oxygen.

George Low is widely credited with having saved the space program in the wake of the *Apollo 1* tragedy. One of the ways he did this was by instilling a renewed sense of pride and purpose in the demoralized teams assembled for the Apollo missions. "No detail was too small to consider," Low says in Deke Slayton and Alan Shepard's book, *Moon Shot.* "We asked questions, received answers, asked more questions. . . . If we made a mistake, it was not because of any lack of candor between NASA and contractor or between engineer and astronaut; it was only because we were not smart enough to ask the right questions. Every question was answered, every failure understood, every problem solved."

Grissom, White, and Chaffee were deeply mourned by NASA members and by everyone connected with the space program, but most of those involved were comforted by the fact that the astronauts had not died in vain. In the *NOVA* special "To The Moon," NASA Director of Launch Operations Rocco Petrone explained why: "If that [fire] had happened while we were on the way to the moon, we'd have lost a crew, never heard from them again, and

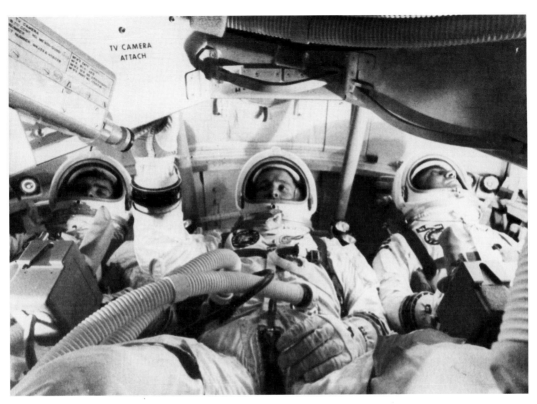

TV CAMERA
ATTACH

there would've been just a mystery hanging over the whole program, which would've caused an untold delay, and maybe even a cancellation." Since the men had died on the ground, where the capsule and all its flaws could be carefully examined, NASA's redesign of the capsule would include safety features to prevent the problems known to have caused the fire.

During redesign NASA also rolled out a new rocket to the launchpad—Wernher von Braun's colossal Saturn V. For the Mercury and Gemini programs, NASA had reconfigured two military rockets, the Redstone and the Atlas, whose original intent was to carry warheads. To get man to the moon, however, NASA needed not only a more powerful rocket, but also one designed specifically for this purpose.

The Saturn V was an engineering marvel, made up of

Apollo 1 astronauts (left to right) Ed White, Roger Chaffee, and Gus Grissom are strapped into the command module during a test a few weeks before their scheduled flight. The day after this photo was taken, a tragic fire broke out in the command module, killing the three astronauts and setting back the Apollo program until safety modifications could be made to the capsule.

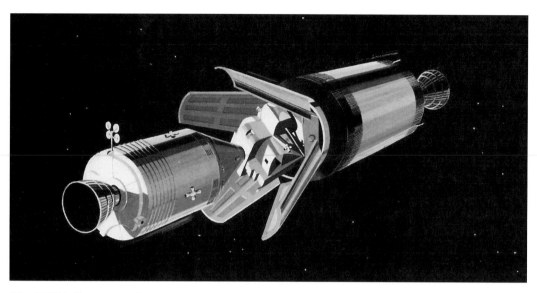

This NASA illustration shows how the command service module, after discarding the third stage of the Saturn V rocket, would turn, dock with the lunar module, and extract it from the expended stage. After this, the spacecraft would turn and fly toward the moon.

several sections designed to perform different functions. Mounted atop the towering rocket was the tiny, 11-foot-high command module (CM). In this cone-shaped section the astronauts would spend most of their mission time; it was the only part of the vehicle that would return to Earth from space. The CM was encased in heat shields that would protect crews from the fiery temperatures—more than 5,000°F—that the craft would generate during reentry into the earth's atmosphere. As the CM sped back to Earth, three huge landing parachutes stowed in its nose would open to slow the capsule's descent before it splashed down in a designated area of ocean. From there, recovery crews would airlift the astronauts to a waiting ship.

Connected to the CM was a 25-foot, cylinder-shaped service module (SM). Without question, the SM was the heart and soul of the spacecraft. It consisted of six interior compartments that contained consumables (fuel, oxygen, water, and batteries that could not be replaced when used) to power the mission. Fuel and oxidizer tanks in the SM gave the service propulsion engine the 20,500 pounds of thrust it needed to fly to the moon, achieve lunar orbit, and return to Earth. The reaction control systems in the SM gave the command service module (CSM) its in-flight steering control. Inside an area referred to as Bay 4 of the SM were two liquid-hydrogen tanks, two liquid-oxygen tanks, and the three fuel cells that combined the hydrogen and oxygen to generate electrical power and produce water. Mounted at the rear of the SM was a high-gain antenna that allowed the astronauts to communicate with Mission Control.

The lunar excursion module (LM), the vehicle that would actually land on the moon, was designed for flying outside Earth's atmosphere—specifically, to transport astronauts to and from the moon's surface. Built by the Grumman Engineering Corporation, the LM was a squat, four-legged craft that looked like an abstract metal sculpture of an enormous insect. Because the LM was not made to function on Earth, its design was not limited by usual engineering constraints such as gravity. Thus, although it was odd-looking, it was a perfectly functional separate spacecraft.

Just like the CSM, however, the LM required life-support systems, and so it was equipped with its own fuel, water, oxygen, and battery supplies. To save on weight, the LM's interior contained the barest essentials. The crew compartment accommodated two astronauts, but there were no seats; the astronauts had to stand during the trip to and from the moon's surface. Two triangular windows provided views of the lunar surface during descent and touchdown; the forward hatch gave the astronauts access to the LM's "front porch" and to the ladder they needed to step down onto the lunar surface.

Below the Saturn V's spacecraft were three-stage engines. The first stage, the largest and bottommost engine, was the S-IC. It consisted of five F-1 engines that delivered 7.5 million pounds of thrust—equivalent to 180 million horsepower, or enough power, according to Wernher von Braun, to "send one fully-loaded DC-3 airliner all the way around the sun." The first-stage engine would lift the 36-story, 6 million-pound Saturn V from its launchpad.

The S-IC would take the vehicle 36 miles into the air, traveling at close to 6,000 miles per hour. Though the S-IC's function was important, it was also brief. After it finished its job, it dropped away, tumbling earthward only 2.5 minutes into the flight.

The second-stage engine, the S-II, picked up where the S-IC left off: its five J-2 engines, smaller than those of the S-IC, would ignite and burn liquid-hydrogen fuel for 6.5 minutes after stage one jettisoned. During this period the vehicle would climb to 108 miles above Earth, moving at 17,500 miles per hour, the velocity necessary to achieve orbit.

Finally, the third-stage engine, a single J-2 engine called the S-IVB, was ignited for approximately two minutes. The S-IVB served two functions: Its initial burn positioned the spacecraft in the proper "parking orbit" around Earth, and later, when the timing and positioning were right, the S-IVB booster was reignited for 6.5 minutes to enable the craft to reach 24,900 miles per hour—the "escape velocity" needed to push the spacecraft out of Earth's orbit and on a course for the moon. In all, the rocket fuel tanks that had taken mechanics 10 hours to fill took just 11 minutes to burn. In approximately 18 minutes the work of the three-part Saturn V rocket was complete.

At launch the Apollo spacecraft perched atop the Saturn V stood 82 feet high and was comprised of the CM, the SM, the LM, the Spacecraft-Lunar Module Adapter (which protected the LM), and a Launch Escape System (LES). The combined command and service modules and the lunar module consisted of 17 tons of aluminum, steel, copper, titanium, and synthetic materials; 33 tons of propellants; 40 miles of wiring; and 1,088 switches and circuit breakers. The entire vehicle, standing 363 feet tall, was made up of more than 6 million parts—all of which had to work perfectly to achieve a successful trip to the moon.

As imposing as the Saturn V looked, no one knew for sure whether this monstrous creation would work. With power equivalent to that of a small nuclear bomb, it could inflict untold destruction on the cape and surrounding areas if it were to explode or otherwise malfunction. Several massive mechanisms had to work in unison within fractions of seconds just to get the rocket off the ground. For example, the nine huge service arms on the steel gantry of the launchpad, each weighing 20 to 30 tons, were connected to the rocket. Four of these functioned as "hold-down arms," to keep the rocket in place after firing until the proper thrust was achieved (within 50 milliseconds). Without them, the Saturn V would topple. As the rocket began its ascent, the five remaining "swing arms" had to retract with equal precision.

The first unmanned Saturn V was launched on November 9, 1967, less than 10 months after the *Apollo 1* fire. It lifted off perfectly. "To me it was the opening of the space age," Launch Control leader Rocco Petrone later said. "Once we had that bird launched, then it was just a matter of time until

Wernher von Braun stands in front of the Saturn V rocket. The 363-foot-tall three-stage rocket provided enough power to shoot the Apollo astronauts out of Earth's gravitational pull and toward the moon.

we got to the moon." Four other unmanned Apollo flights followed, and in the fall of 1968 the first manned Apollo mission was ready for launch.

Wally Schirra, Walter Cunningham, and Donn Eisele were in *Apollo 7* as it lifted off the cape on October 11, 1968, bound for an 11-day, 163-orbit flight around the earth. The machinery worked exactly as it was meant to. America's race to reach the moon was back on track.

While the United States was perfecting its Apollo program, the Soviet Union was also experiencing the ups and downs of space exploration. Three months after the *Apollo 1* tragedy, cosmonaut Vladimir Komarov was killed when his *Soyuz 1* spacecraft spun out of control during reentry and plunged to the ground. Four days after *Apollo 7* returned to Earth, however, the Soviets managed to launch two separate crafts (*Soyuz 2* and *Soyuz 3*) a day apart from each other—one was unmanned, the other carried cosmonaut Georgi Beregovoi. The cosmonaut maneuvered *Soyuz 3* within feet of *Soyuz 2* and broadcast images of his accomplishment to his countrymen on Earth. Though the two ships never docked, the U.S.-Soviet race to the moon was entering the backstretch. Who would finish first was anybody's guess.

As the race heated up, NASA decided to reformulate the mission goals and crew of *Apollo 8*. Originally, James McDivitt, David Scott, and Russell Schweickart were assigned to the *Apollo 8* mission to test the CM and LM during Earth orbit. Some months earlier George Low, Bob Gilruth (director of Houston's Manned Space Center), and Chris Kraft (director of Houston's NASA Flight Operations) had discussed sending the next mission, *Apollo 9*, into a circumlunar orbit, marking the first trip by U.S. astronauts out of Earth's orbit and into orbit around the moon.

But problems with the LM's construction and design pushed back the delivery date of the craft by several months. Since McDivitt and his crew had been training for some time to complete the CM and LM tests, Deke Slayton decided that perhaps the crew in training to actually orbit the moon should switch missions with McDivitt's team. He put in a call to Frank Borman, commander of the *Apollo 9* mission. In *Moon Shot*, Slayton recalls Borman's reaction to their conversation: "[He] almost turned handsprings when I suggested . . . that there was a possibility [his crew] would go all the way to the moon. His answer was an overwhelming yes."

Time seemed to be running short for America to

achieve the first manned lunar orbit. The government's Central Intelligence Agency (CIA) had received reports that the new Soviet spacecraft, the *Zond*, was being readied to take a cosmonaut around the moon as early as December 1968 or January 1969. As it turned out, the Soviets hadn't advanced quite that far—but NASA was taking no chances. On December 21, 1968, Borman and crewmates Jim Lovell and Bill Anders achieved a flawless liftoff from Cape Canaveral. Three hours later Mission Control radioed the crew: "*Apollo 8*, you are go for TLI." The command allowed Borman, Lovell, and Anders to begin their translunar injection—the maneuver by which an additional engine burn would thrust them out of Earth's orbit and point them toward the moon. Until this moment, man's farthest reach was 850 miles from Earth. *Apollo 8* would shatter that record, traveling 240,000 miles from the planet.

Three days later, on Christmas Eve, the crew of *Apollo 8* broadcast a splendid gift to mankind: the first close-ups of the moon's surface. They also described for the first time what the "far side" of the moon looked like. (Because of the way the moon orbits Earth, one hemisphere remains obscured from Earth and is in constant darkness.) That side, the astronauts said, was much more rugged-looking than the hemisphere seen from Earth. Acting as a lunar tour guide, Jim Lovell described the moon's surface as resembling plaster of Paris or "grayish beach sand." Finally, Lovell added, "The vast loneliness is awe-inspiring, and it makes you realize just what you have back there on Earth."

But the awe and excitement with which the astronauts viewed the moon was quickly overpowered by the magnificent vision of Earth as it rose above the horizon of the moon (much like the sun rises above Earth's horizon). Suddenly, the vast monochromatic view through the capsule windows was splashed with brilliant, shimmering color—the colors of Earth, a living, fragile-looking planet.

Borman later said that the first view of Earth from space was "the most beautiful, heart-catching sight" of his life.

Before signing off from their Christmas Eve broadcast, the three astronauts read a passage from the biblical Book of Genesis to the largest audience in the history of mankind. In "To the Moon," NASA Flight Director Gene Kranz described the moment: "There's a lot of times in my life when I've been brought to tears by just the power, the immensity, the beauty of what we were doing, and this was one of those days."

Borman, Lovell, and Anders finished their in-flight chores, which included checking instruments and taking photographs of potential lunar landing sites—and of that first glorious look at Earth from 240,000 miles away. Then the astronauts successfully fired the engines to take them out of lunar orbit. *Apollo 8* splashed down in the Pacific Ocean near Christmas Island on December 27, 1968.

The mission seemed to have operated under a charm: there were no glitches or problems, and the astronauts received a hero's welcome when they arrived home. Two weeks later, in an address to Congress, Borman emphasized how vitally important the space program was to America's development. "Exploration really is the essence of the human spirit," he said. "And to pause, to falter, to turn our back on the quest for knowledge is to perish."

The *Apollo 9* mission tested moon-landing hardware, including Grumman Engineering's newly completed LM; *Apollo 10* repeated *Apollo 8*'s moon orbit flight with a heavier payload, including the LM itself. By mid-May of 1969 NASA was ready for its first manned mission to the surface of the moon. Neil Armstrong would command the mission and would become the first human to walk on the moon. Buzz Aldrin would also travel to the moon while CM pilot Michael Collins orbited in preparation for their rendezvous and trip back to Earth.

On the morning of July 16, as more than a million people crammed into the area surrounding Cape Canaveral,

Apollo 11 roared into history. Four days later *Eagle* gently touched down on the moon, in an area known as the Sea of Tranquility. At 10:56 P.M. EDT that same day, televised coverage showed Armstrong climbing down the LM's ladder, pushing himself away from the craft, and planting both feet on the moon. Around the world people were awestruck as they heard the first words beamed back from the moon's surface: "That's one small step for man, one giant leap for mankind."

About 15 minutes later Aldrin stepped onto the moon as well. Armstrong and Aldrin gathered moon samples, set up three experiments, and took scores of photographs. Before they boarded the LM, the two men on the moon unveiled a commemorative plaque of their historic visit and planted an American flag. The race to the moon was over. The United States had triumphed.

The United States won the race to the moon, as Apollo 11*'s lunar module* Eagle *touched down in the Sea of Tranquility on July 20, 1969. This photo of astronaut Edwin "Buzz" Aldrin with the American flag was taken by Neil Armstrong, the commander of the* Apollo 11 *mission.*

James A. Lovell Jr., in a photo taken a few years after he was accepted into the astronaut corps for Project Gemini. When he was selected as commander of Apollo 13, *Lovell had logged more than 570 hours in space.*

4

THE PINNACLE OF AN ASTRONAUT'S CAREER

"You guys are pretty good at measuring bilirubin, but the one thing you never thought to measure is persistence and motivation."
—Jim Lovell, addressing the doctor who rejected him from the space program in 1958

From the time Jim Lovell was a young man, he had dreamed of rockets and space flight. At 17, while visiting his aunt in Oak Park, Illinois, Lovell hopped a train into downtown Chicago on an expedition to buy some chemicals. He even had a list prepared—potassium nitrate, sulfur, and charcoal. He hoped to buy amounts of these substances to take back to Milwaukee, where he and two friends were going to build a working rocket. Initially, they had wanted to build a liquid-fuel rocket along the lines of those developed by pioneers Robert Goddard, Hermann Oberth, and Wernher von Braun—but such plans proved beyond their capabilities, so the three teens had settled on a solid-fuel rocket instead.

An aerial view of the U.S. Naval Academy at Annapolis, Maryland, taken around the time Jim Lovell (inset) attended the school. After graduating from the academy, Lovell was selected for training as a fighter pilot.

What Lovell didn't realize at the time was that he was essentially asking for the components of gunpowder. He left the Chicago chemical factory empty-handed but not discouraged. With the help of their chemistry teacher, the boys fashioned a makeshift rocket from cardboard tubing, wood, and a drinking straw "fuse" filled with chemicals. On launch day they packed the bottom of the rocket with homemade gunpowder, planted it in the middle of an empty field, lit the fuse, and scrambled for cover. To their amazement, the rocket sputtered and lifted about eight feet off the ground, zigzagged through the air for a moment, and then exploded gloriously. For the other boys, the experiment was a brief moment of good-natured fun. For Jim Lovell, however, it was a milestone.

Lovell had become interested in rockets years before, when he stumbled across some books about basic rocketry. Even before he and his friends successfully launched their craft that Saturday afternoon, he had decided to make it his career. While in high school he applied to the U.S. Naval Academy in Annapolis, Maryland. But he was selected only as a third alternate, so he signed on for a stint in the navy's Holloway Plan, a recruitment program that offered a paid college education and naval aviation training. After two years of studies Lovell reported for flight training and six months of active duty at sea. He applied to Annapolis again, and this time he was accepted. Attending the naval academy meant that Lovell would be demoted from novice aviator to plebe, but he didn't mind. He believed that the academy offered him the best education available, and besides, he could apply to naval flight school again after he graduated.

As it turned out, Lovell had made a fortunate choice. Shortly after he decided to attend Annapolis, the United States entered the armed conflict in Korea. Most of Holloway's aviators-in-training were pressed into war service, and those who survived were unable to resume training until seven years later. Had Lovell stayed with the Holloway program, his educational career would have suffered the same delays.

Just hours after Midshipman James A. Lovell Jr. graduated from the U.S. Naval Academy in the spring of 1952, he married his hometown girlfriend, Marilyn Gerlach, in St. Anne's Episcopal Cathedral in downtown Annapolis. Out of 783 graduates, Lovell was one of 50 selected to continue with naval flight school. Fourteen months later, his training completed, Lovell was assigned to Moffett Field in California, where he joined the Composite Squadron Three, a group of aviators who had the unenviable job of night flying on aircraft carriers.

Taking off from and landing on a carrier deck that from 1,000 feet in the air resembles a bathroom tile was one of the most difficult flight assignments Lovell could have drawn. The maneuver required steely nerves and excellent flying skills. Eventually Lovell achieved 107 successful landings, and after some time as a flight instructor for other pilots, the 29-year-old sought an even more challenging assignment. He found just such a project at the U.S. Aircraft Test Center, located near the naval air station in Patuxent River, Maryland.

Lovell figured that his new job—test-piloting experimental military aircraft—was about as close as he was ever going to get to space flight. But shortly before he and Marilyn .noved their growing family (a three-year-old daughter, a two-year-old son, and a third child on the way), startling news spread across the country. The Soviet Union had put a satellite into orbit above the earth. No matter, Lovell thought as he pulled his packed Chevy out of the driveway for the long journey from California to Maryland. His involvement in aeronautics was confined to test-flying planes for the military, so he needn't worry about satellites in space. Lovell didn't realize that the satellite, dubbed *Sputnik 1* by the Russians, had at last galvanized the United States into launching its own space program.

Lovell was disappointed to discover that he had been assigned to the electronics test (radar testing) division after completing his flight training, but that project was about to change as well. The fledgling NASA program, based in Langley, Virginia, was about to launch Project Mercury, and officials had begun working on a list of astronaut candidates, drawing on officers in the air force and navy who met their criteria. Jim Lovell was one of the first 34 (out of 110) men to be invited to participate in the program; so were his Pax River colleagues Pete Conrad and Wally Schirra. Lovell was dumbstruck. Was it possible that he was about to take the final step toward his dream of going into space?

The answer—after a week of medical tests in which Lovell's body was endlessly poked, stuck, probed, punctured, and pumped—appeared to be no. The officiating doctor blamed his rejection on a slightly high level of bilirubin (a natural liver pigment). In *Lost Moon*, Lovell recounts his attempts at persuading the doctor to reconsider. "If you only accept perfect specimens, sir," he cleverly argued, "you'll only wind up with one kind of data. Taking someone with a little anomaly means you'll learn even more."

The reasoning didn't work; Lovell reported back to Pax River the next day. Shortly after him came Pete Conrad, also rejected (although Lovell's other friend, Wally Schirra, was accepted). A few weeks later Lovell and Conrad watched on television as NASA introduced the seven Project Mercury astronauts. During the next three years Lovell watched, rather than participated in, the Mercury missions for which he had been told he was unfit.

Meanwhile, his career at Pax River was going well. After the electronics test and armaments test divisions merged, Lovell ended up in weapons testing. His first assignment was to evaluate the new F4H Phantom, an aircraft specially designed to fly night combat missions. Lovell worked with the employees of McDonnell Aircraft, and when the F4H was ready for its first test flight, he was named flight director. The new assignment meant another relocation, this time to the Oceana Naval Air Station in Virginia Beach, Virginia.

By the summer of 1962 the Mercury space program was winding down; only Schirra and L. Gordon Cooper had yet to fly. One afternoon while Lovell sat in the ready room at Oceana, he picked up a copy of *Aviation Week and Space Technology* and idly flipped through its pages. Then he suddenly zeroed in on a small news piece. The report discussed a recent NASA press release in which the

Jim Lovell pedals an exercise bike as Dr. Charles A. Berry, the chief of NASA's medical program, examines results. This photo was taken as Lovell was preparing for his Gemini 7 *mission in 1965. Seven years earlier, NASA doctors had prevented Lovell from becoming one of the seven original Mercury astronauts.*

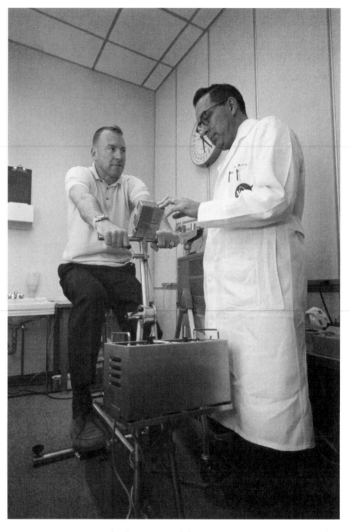

agency announced plans to add 5 to 10 astronauts to the Manned Space Flight Program.

Lovell had many reasons simply to forget this whole space business. He was 34 years old—a relatively advanced age for the kind of work such a project required. Moreover, he had already been rejected once. And this time around, he was certain, the number of applicants would soar. Despite the odds, however, Lovell decided to reapply. As it turned out, the second time was the charm.

A little more than two weeks after his trip to NASA to interview for the Gemini program, Lovell received a call from Deke Slayton, the director of NASA Flight Crew Operations in Houston. Slayton told Lovell that he was calling regarding NASA's selection of the next group of astronauts. Would he like to come work for NASA? Jim Lovell enthusiastically shouted, "Yes, yes! Of course!" Then he hung up and let the news sink in. He was going to Houston. There he might just get the opportunity of a lifetime: the chance to sit atop a rocket as it blasted off for the moon.

From that time on, there was no stopping Jim Lovell. By the spring of 1970 he had participated in three space missions, two as a Gemini astronaut and one as a member of the *Apollo 8* space crew—the first men to orbit the moon and view its far side. All that remained for him now was the crowning achievement of a storybook aeronautical career: walking on the moon itself. And *Apollo 13* would allow him to reach that long-sought-after goal.

Command service module 109, which would become part of Apollo 13, *hangs in a clean room before being lifted to its position atop the Saturn V. No one realized that there was a damaged part deep within the service module that would disrupt the mission.*

5

"HOUSTON, WE'VE HAD A PROBLEM"

"The ground may not have believed what it was seeing, but we did. It's like blowing a fuse in a house—the loss is a lot more real if you're in it. Things turn off. We believed that the oxygen situation was disastrous, because we could see it venting."

—LM pilot Fred Haise, shortly after the
explosion aboard *Apollo 13*

BY THE TIME JIM LOVELL and his *Apollo 13* crewmates followed the triumphant journey of *Apollo 11* and the successful lunar science expedition of *Apollo 12*, Lovell was one of the most seasoned astronauts in the NASA program. He had logged 572 hours in space and had flown nearly 7 million miles, making him the most traveled man on the planet.

Such statistics would have been more than enough to satisfy many NASA career men. But not Lovell. He hadn't yet reached his highest goal: he wanted to go to the moon. Even more important than landing on the moon, however, was being assigned as commander of a lunar mission. With *Apollo 13* it seemed Lovell would get the chance to do both.

When the original Apollo mission flight and crew selections were handed out, Jim Lovell was tapped as commander of the backup crew for the *Apollo 11* voyage. After the mission's outstanding success, Lovell, Haise, and Ken Mattingly began training for their own moon mission, *Apollo 14*, scheduled for October 1970. But *Freedom 7* hero Alan Shepard had other ideas.

Shepard had been grounded since his historic 15-minute Mercury flight because of an inner ear imbalance. Anxious to regain active flight status, he underwent a surgical procedure to correct the problem, and now he was out to snag himself a lunar mission assignment. He discussed the possibility with his old friend Deke Slayton, and though NASA was willing to send Shepard on a moon mission, they knew he was a bit rusty from his layoff. Thus, despite Shepard's eagerness, Slayton asked Lovell to switch missions with the former Mercury astronaut so that Shepard could fly with *Apollo 14*.

Lovell agreed without a moment's hesitation. He couldn't wait to get back into lunar orbit and was more than happy to go six months earlier than originally scheduled. The motto he chose for the *Apollo 13* mission was "*Ex luna, scientia*," a Latin phrase meaning "From the moon, knowledge." He named his LM *Aquarius*, after an ancient Egyptian mythological water carrier who brought prosperity to the Nile valley (and for whom a constellation was named). Lovell called the CM *Odyssey* simply because he liked the sound of the name. Perhaps he was unaware of the standard definition of the word: "a long voyage marked by many changes of fortune." But that knowledge probably wouldn't have bothered Lovell even if he had known. Hadn't he advanced through most of his career by changes of fortune, after all?

Joining Lovell on this mission was rookie LM pilot Fred Haise, 36. A native of Biloxi, Mississippi, Haise

Jim Lovell (left) films a rock sample held by Fred Haise during one of their geological field trips.

became a naval aviation cadet in 1952 and was stationed at the naval air station in Pensacola, Florida. His military career included stints as a pilot with the Oklahoma National Guard's 185th Fighter Interceptor Squadron and the 164th Tactical Fighter Squadron in Mansfield, Ohio. After graduating with honors and a B.S. in aeronautical engineering from the University of Oklahoma in 1959, Haise, his wife, Mary, and their two children moved to California, where Haise began working as a test pilot at NASA's Flight Research Center.

Haise was one of the 19 astronauts selected by NASA in 1966 for its space program. His love for the exploration and science of space travel matched Lovell's fascination with rocketry. To prepare for their lunar mission, the two took geological field trips with California Technical Institute geologist Dr. Leon Silver that included visits to Meteor Crater, Arizona, and to basalt fields in Iceland. The trips helped Lovell and Haise learn how to identify various kinds of rock, lava, and soil samples. Haise was especially drawn to the studies, and he was dubbed "the drilling fool," after the extraordinary pleasure he derived from practicing with the geological equipment designed for lunar use.

In the early days Haise might not have fit NASA's astronaut profile the way Alan Shepard, John Glenn, or L. Gordon Cooper had. But as the focus of the space program shifted from test-piloting to exploration, Haise turned out to be precisely what NASA needed. He had an excellent reputation among his fellow astronauts, and his dedication to the project was well known among contractors like Tom Kelly, the engineering manager for LM designer Grumman Engineering.

While many of NASA's senior astronauts were busy training for missions, members of the "third string"—those among the original 19 who were chosen for the program but not named for missions—immersed themselves in studying the hardware that would take their colleagues to the moon. Pilots like Ken Mattingly and Jack Swigert spent most of their time in Downey, California, where the CM was undergoing extensive testing. Haise, meanwhile, was assigned to Grumman, where he inhabited the LM almost around the clock. Andrew Chaikin, in his book *A Man on the Moon*, describes Haise's unerring commitment to the Apollo program:

> Haise all but lived at the [Grumman] factory. He developed the kind of intimacy with the craft that only a "company pilot" gets. He knew far more about the guts of the lander than anyone needed to know in order to fly it. . . . He knew just about every one of the odd, one-of-a-kind parts that Grumman fashioned at the height of the weight-saving crunch. More nights than he could count, Haise had curled up on the floor of a lunar module and catnapped while a test was delayed. After two years of this he felt he knew the machine almost like a part of himself.

Ken Mattingly, another member of "the 19," was slated to serve as command module pilot for the *Apollo 13* mission. Like Haise, 34-year-old Mattingly had spent countless hours learning everything he could about his vessel. Tall, quiet, and unassuming, Mattingly at first glance did not seem to fit the profile of a rugged naval aviator who had flown aircraft from the carriers *Saratoga* and *Franklin D. Roosevelt*. While Lovell and Haise were off on geological field tests, Mattingly studied photography, learning how to take photographs of the moon from the orbiting command module, where he would be while Lovell and Haise were on the surface.

In addition to getting specialized training, the crew of *Apollo 13* also trained together constantly in the nine months before their scheduled flight. They worked so closely for so long that they developed the ability to read even the subtlest nuances in one another's voices. All seemed to be going well for what was jokingly called the "Lucky 13" crew—until the weekend before launch. Backup LM pilot Charles Duke Jr. came down with a case of German measles, and NASA doctors were concerned that he might have passed the condition to other people with whom he had been in contact during the disease's two-week incubation period. These people included *Apollo 13*'s prime crew—Lovell, Haise, and Mattingly. By Wednesday blood tests from the three showed that Lovell and Haise were probably immune, but Mattingly's tests were inconclusive, and doctors could not be certain that he would not come down with measles. They believed they had no choice but to pull Mattingly from the mission and replace him with his backup, Jack Swigert.

Deeply disappointed, Mattingly nevertheless continued to feel fine and showed no signs of getting sick. Meanwhile, however, Swigert practically began living in the CM simulator in a last-minute rush to prepare for the upcoming mission. With just a few days of training time

Thomas K. "Ken" Mattingly, the CSM pilot scheduled to fly on Apollo 13*, was pulled from the mission after he was exposed to measles. NASA did not want to take the risk of the pilot getting sick while in space. As it turned out, Mattingly never developed any illness.*

together, Lovell and Haise began feeling comfortable about the replacement only the day before the launch, when NASA doctors finalized their decision: Mattingly was officially out; Swigert was in.

Swigert was NASA's lone bachelor astronaut. The former test pilot had a reputation for leading a wild, carefree life with the same enthusiasm with which he pursued his passion for flying. A swank dresser, Swigert was as fastidious in his work as he was in his appearance: he was one of the many men who had worked tirelessly at the tedious job of rewriting checklists and revamping procedures after the fire on *Apollo 1*. And like Haise and Mattingly, he spent

John L. "Jack" Swigert, the backup pilot for Apollo 13, *replaced Ken Mattingly just days before the rocket was supposed to blast off. Swigert underwent several days of intensive training in the command module simulator with the other members of the crew until everyone felt comfortable working together.*

nearly all of his time at North American Aviation, the builder of the command module, while the vessel underwent testing. He did differ from his crewmates in one way, however. Whereas many astronauts seek to be one of the two men in a lunar mission who step onto the moon from the LM, Swigert had approached Deke Slayton with a request that he be assigned to pilot a Command Module.

The intense disappointment that Lovell and Haise felt about Mattingly's being bumped from the team took the edge off some of their excitement. Swigert, on the other hand, had been on a roller coaster of emotion. In just one week's time he had gone from being backup pilot and concierge for the prime crew to command module pilot on his way to the moon.

Apollo 13, officially known by the unwieldy name of NASA Spacecraft 109/Lunar Module 7/Saturn 508, lifted off Cape Canaveral on schedule at 1:13 P.M. Central Standard Time (CST) under bright, clear skies on April 11, 1970. The launch was nearly perfect; only the slight vibration during the rocket's ascent and the premature cutoff of the S-II booster engine prevented the liftoff from being routine.

By now—amazingly—launches had become relatively uneventful for Americans, who had witnessed not only the thrilling early launches of Project Mercury, but also the series of Gemini launches and several Apollo missions. As a result, the early, intense interest in the lunar missions had begun to wane by the time *Apollo 13* roared up into the sky. Mission Control Lead Flight Director Eugene Kranz explained what he believes are the reasons behind such a phenomenon:

> [Loss of interest in the program] wasn't unexpected. If you really take a look at it, it was a very tumultuous era. We assassinated three public figures [John F. Kennedy, Robert F. Kennedy, and Martin Luther King Jr.]. We had the war in Vietnam. We had all the peace demonstrations. We had Kent State. We had every anarchist in the world hijacking an airplane and taking it somewhere and blowing it up. There was a lot of competition for the headlines in those days.

Three hours into the mission, *Apollo 13* blasted out of Earth's orbit to begin its long journey to the moon. For Lovell, now on his fourth mission, the trip was almost business as usual. But for Haise and Swigert every stage of the flight was a first, and they tried to soak up each memorable moment. At 5:14 P.M., four hours after takeoff, Swigert completed the transposition maneuver, pulling the command service module away from the third-stage

S-IVB engine, and then turning it around to dock with the LM (*Aquarius*) and extracting it from atop the S-IVB. Moments later the ground crew began the first scientific exercise of the mission by setting the S-IVB on a collision course with the moon. The seismometer left on the surface by the *Apollo 12* crew would relay information from the impact to Earth for later study.

Aquarius was scheduled to land on a hilly region of the moon dubbed Fra Mauro, 112 miles east of the site where *Apollo 12*'s LM, *Intrepid*, had landed. The soil and rock samples collected by astronauts Pete Conrad and Alan Bean during that mission were much like those brought back by Neil Armstrong and Buzz Aldrin from the *Apollo 11* landing. Scientists were now hoping that the hilly Fra Mauro area would yield different samples, which might shed light on the ages of the varying lunar topography.

The choice of landing sites and the time of day during which Lovell and Haise would touch down made *Apollo 13*'s route rather perilous. The ship would not be able to follow the usual free-return trajectory, a course that sends the spacecraft behind the moon and then out again using the moon's gravitational pull. Instead, the crew would need to adjust the orbit manually. At 7:53 on Sunday evening Swigert fired the engines and successfully moved the linked *Odyssey* and *Aquarius* out of their free-return trajectory.

Marilyn Lovell remembers thinking that her husband looked and sounded very relaxed during his Monday-evening televised broadcast, considering the circumstances. In the broadcast Swigert stayed mostly in the background, while Haise floated between the two joined craft as Lovell followed him with the camera. As the broadcast wound down, Lovell shared his fascination with a portable tape recorder he had brought with him. First he released his grip on the machine and let it float to demonstrate the weightless atmosphere. All the while, the recorder blared music from the soundtrack to the movie

2001: A Space Odyssey. Shortly after, Lovell signed off: "This is the crew of *Apollo 13*, wishing everyone a nice evening. We're just about ready to close out *Aquarius* and get back for a pleasant evening in *Odyssey*. Good night."

At NASA the projection screen went blank. Mission Control was still communicating instructions to the flight crew as Marilyn Lovell and her daughters and Mary Haise prepared to leave. Marilyn hoped to get home in time to tuck their youngest son, Jeffrey, into bed. Though she enjoyed seeing her husband on the broadcast from space, she hadn't particularly wanted to make the 15-minute drive to the Manned Spaceflight Center from their home in the Houston suburb of Timber Cove. She felt that her husband had been somehow cheated of his "moment" on the air, and she was angry with the networks and annoyed with NASA. She could not know that in just a short time an unforeseen event would place her husband and his crewmates at the center of nearly continuous news coverage around the world—from the same networks that had deemed the mission of little public interest only an hour earlier.

As Haise finished closing up *Aquarius*, EECOM Sy Liebergot waited for Swigert to complete the requested cryo stir of the oxygen and hydrogen tanks. As Swigert turned on the fans in the tanks, a yellow warning light flashed on his CM console. On the ground, Liebergot saw a similar light on his own console, and he guessed that it signaled a problem with the pressure in the cryogenic system. His assumption seemed confirmed when he scanned his screen and noticed a low pressure reading in one of the hydrogen tanks—a persistent problem that he had monitored for the past two days.

Then came the sudden bang that shook the spacecraft and startled the crew. If it was not Haise playing tricks on Lovell and Swigert, then what in the world could it be? Almost immediately warning lights began flashing on instrument panels and alarms sounded in the astronauts' headsets. Swigert noticed an immediate loss of power in

main B bus, one of two power distribution panels that provided electricity for the CM.

"Hey, we've got a problem here," Swigert shouted into his microphone. On the ground, Capcom Jack Lousma was the first to respond: "This is Houston, say again, please." Lovell answered: "Houston, we've had a problem," he said with surprising calm. "We've had a main B bus undervolt."

Lovell's words sent jolts of adrenaline through every member of the Mission Control crew. Liebergot began scanning his consoles for a clue to what had just happened on the spacecraft. He couldn't believe what he was seeing. The pressure level in the second oxygen tank had dropped to zero. Then he noticed that fuel cell number one was also dead. If this information was correct, Liebergot realized, *Odyssey* had suddenly and inexplicably lost most of its

Eugene "Gene" Kranz, in the Mission Operations Control Room, reacts to the problems on Apollo 13. *As lead flight director, Kranz was in charge of the effort to bring the astronauts home safely.*

power and half of its life-support systems. It just didn't make sense.

While controllers on the ground were checking their data, Lovell and Swigert sealed the tunnel between *Odyssey* and *Aquarius*. They suspected that a meteor had struck the LM, and they wanted to prevent more oxygen from leaking into space. Lovell's heart sank. If this problem wasn't just an instrument panel on the fritz, then the mission was lost—and with it, his last chance to place his own boot prints on the powdery surface of the moon.

The Mission Control crews operated in shifts, whose teams were named for colors. The White Team had been on shift since 2:00 that afternoon; at the flight director's console was 10-year NASA veteran Eugene Kranz, the White Team's leader. He had begun receiving malfunction reports in his earpiece from several members of his team. Among them was the report from Instrumentation and Communications (INCO) that *Odyssey*'s radio had quit transmitting from its single high-gain antenna and had automatically switched over to its four omnidirectional antennae, located at the base of the service module. Kranz and his team were looking at so many failures that he believed the problem had to be in the instrumentation and not in the spacecraft itself.

Up in *Odyssey*, however, Lovell, Haise, and Swigert could see the command module dying before their eyes. The bang had caused the craft to begin swaying and wobbling; Lovell struggled to stabilize it, but he was having no luck. In frustration Lovell unbuckled himself from his couch and floated to one of *Odyssey*'s windows in the hope of gaining information about the cause of the problem. What he saw outside the spacecraft prickled the hair on the back of his neck.

The ship was enveloped in a milky, gaseous cloud. Something was "bleeding" from the ship into space. "It looks to me that we are venting something," he announced. "We are venting something into space. It's a gas of some sort."

Liebergot began to put the pieces of this puzzle together. If the tank readings were zero and Lovell was seeing gas venting from the craft, then it had to be an oxygen leak. Kranz, realizing that the problem was growing more serious with each minute, worked to keep everybody composed. "Okay," he said. "Let's everybody keep cool. Let's make sure we don't do anything that's going to blow our electrical power or cause us to lose fuel cell number two. Let's solve the problem, but let's not make it any worse by guessing."

Lovell, Haise, and Swigert began to understand the enormous impact of the problem. If the instrument readings were accurate—and Lovell was pretty sure they were—then in a short time all the main power and oxygen would drain from *Odyssey*, leaving the crew to survive on only the tiny batteries and the oxygen stored for reentry in the command module. Those supplies would last just a few hours.

But *Odyssey* had a "lifeboat"—*Aquarius*. As a separate craft, *Aquarius* was equipped with its own power supply and life-support system. Lovell looked at his shipmates. Finally he said, "If we're going to get home, we're going to have to use *Aquarius*." There seemed to be no other choice.

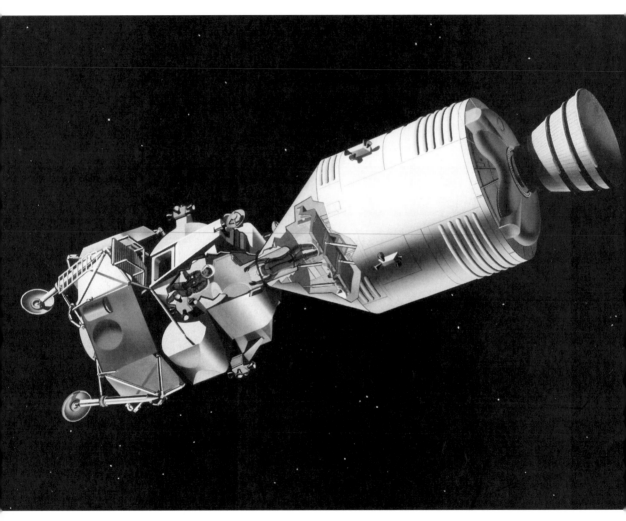

The cutaway in this NASA illustration shows how the astronauts could leave the command module and enter the lunar module through a small tunnel. With Odyssey *out of power, the three astronauts were forced to move into the lunar module* Aquarius *and use its life-support systems to survive during the trip back to Earth, more than 200,000 miles away.*

6

A LIFEBOAT
FOR APOLLO

"So you see, survival now became [a matter of] initiative and ingenuity and was one [goal with] which the ground [crew] continually helped us. . . . We had all kinds of people on the ground trying to think of ways of extending our lifetime."
—*Apollo 13* Commander James A. Lovell

At about the same time Lovell, Haise, and Swigert were considering powering up *Aquarius* and moving into it, Mission Control was reaching the same conclusion. Capcom Jack Lousma heard the urgency of the situation on the communications loop from EECOM. Liebergot and his Black Team replacement, Clint Burton, continued to be the bearers of bad news. It seemed only a matter of minutes before *Odyssey* would cease to function and be unable to keep the astronauts alive.

In the dying CM, Lovell made his own assessment of the consumables. Although getting all three men into *Aquarius* seemed to be the only option, even that plan had one major hitch: the LM had been

73

Former Grumman engineer Tom Kelly, in a 1999 photo, holds a model of the lunar module that he and his team had designed for the space program. During the Apollo 13 *crisis, Kelly and his men had to determine how the lunar module could be used as a lifeboat to save the three astronauts.*

designed to accommodate and support only two men, and only for up to 45 hours. *Apollo 13* was more than 200,000 miles from Earth; Lovell estimated that the journey home would take roughly 100 hours.

As Liebergot sat at his console watching the life drain out of *Odyssey*, his craft counterpart, Bob Heselmeyer, the telemetry, electrical, and EVA mobility unit officer (TELMU) who monitored the lunar module, sat next to him. But Heselmeyer was viewing very different data— good, healthy-looking data was being transmitted to his console screen from *Aquarius*. The situation was stable for now, but that wouldn't last. Heselmeyer, his Mission Control colleagues, and the crew at Grumman led by Tom

Kelly, faced the formidable task of making the LM perform in ways for which it had not been designed.

Finally, Gene Kranz gave the order to Swigert to begin powering down *Odyssey*. They were to shut the two dead reactant valves on the fuel cells, which normally regulated the oxygen flow from the cryogenic tanks to the fuel cells. Once they had been closed off, the valves could not be reopened. Yet Liebergot hoped that doing so would stop the oxygen in tank one from leaking, and he thought it was a risk worth taking.

Since Apollo mission rules required all three fuel cells to be up and operating perfectly for a lunar landing, Haise knew then that the mission was lost. In *Chariots for Apollo*, he describes what he was thinking as he began to realize how grave the astronauts' situation had become: "We were never scared," he says, "only disappointed and sick that we could not land on the moon. We had lost a mission. We had worked a lot of years, and seemingly for nothing. The worst feeling was right away, when I looked at the meter and saw that the oxygen tank was gone, and knew that the landing had to be scrubbed."

But shutting the valves made no difference in the tank's pressure level. With no other options, Lousma finally radioed the flight crew to make their way into *Aquarius*. Haise floated through the tunnel first, saying to Lovell, "I didn't think I'd be back here this soon." Lovell replied, "Just be happy it's here to come back to." He was right: time was running out, and they still needed to complete the lengthy processes of powering up the LM and transferring the guidance system coordinates, necessary for the spacecraft to stay on course, from *Odyssey* to *Aquarius*.

Now the astronauts were in a race against the clock. Lovell struggled with the guidance computations that Swigert called out to him. To be sure their math was correct, they asked Mission Control to check the calculations before Haise entered them into the LM's computer. After that, Haise and Lovell continued to follow the power-up

At Mission Control, a team of astronauts review data from Apollo 13. *They include (left to right) Deke Slayton, the director of flight crew operations; grounded CM pilot Ken Mattingly, who was called in to help during the crisis; capsule communicators Vance Brand and Jack Lousma; and John Young, the backup commander of* Apollo 13.

procedures for *Aquarius*.

Unlike the calm, ordered routine normally followed during an Apollo mission, the power-up procedure was chaotic. Ground controllers called up scores of instructions, and the flight crew called down questions—the communications loop was filled with colliding voices. The confusion produced its share of tense moments: at one point, for example, Glynn Lunney, the flight director on duty, accidentally ordered the attitude-control jets powered down in *Odyssey* before those in *Aquarius* were brought online. For one excruciating moment neither ship had steering. Still, the crew's ability to adapt was a marvel to those at Mission Control. In the few minutes before

Odyssey's oxygen ran out, Haise and Lovell managed to complete a series of procedures that would normally have taken about two hours.

Swigert, still in *Odyssey*, did not leave the ship until the last of the power-down switches had been flipped. He took one last look around. The command module, now dark and lifeless, was going to get very cold very soon. He only hoped that when the time came to power the CM back up, it would actually come back to life long enough to get them home. Having finished his procedures, Swigert now had to wait until it was time to prepare *Odyssey* for reentry. Squeezing in between Lovell and Haise in *Aquarius*, Swigert said to them, "It's up to you now."

The Black Team had just come on shift in Mission Control; their job was an unenviable one. Lunney was faced with a host of ongoing problems, each of which seemed to need his immediate attention. In reality, however, the crisis could be broken down into three phases: powering down the CM while powering up the LM; devising a route that could bring the crippled ship home without depleting its dwindling resources; and insuring a safe reentry and recovery of *Odyssey* and the three astronauts. They had already solved the first problem, to everyone's great relief. Now they needed to figure out how to get the crew back home.

Because of the drain on consumables that kept the crew alive and the LM functioning, several members of the Mission Control team held widely varying views of how they were going to accomplish the task. Gene Kranz listened to all the options, then discussed the situation with his mentor, Chris Kraft. A direct abort (bringing the spacecraft to a complete stop, turning the ships around, and firing the engines to head back to Earth) would bring the crew home in 55 hours. But because Kranz didn't trust the condition of the service module engine, he felt a direct abort was too risky. The other option—a circumlunar abort—would use gravitational pull to send the craft around the moon and put it on a "slingshot" trajectory back to Earth. This method would take longer, but it required fewer demands on the machinery and the astronauts. Kranz decided to go for the second plan, and Kraft concurred.

To complete this maneuver, the astronauts first needed to initiate a short burn of the LM's engine to get the craft back on a free-return trajectory. Then, several hours later, they had to perform a second, tricky burn called "PC + 2." The letters PC stood for "pericynthion," a predetermined point in the mission located on the far side of the moon. Under normal circumstances, the CSM's engine would be fired at this point; this would place the craft into lunar orbit. However, if the engine failed, the LM's engine would be ready to make an emergency burn at the point two hours past pericynthion ("+ 2") to send the spacecraft back to Earth. This had been calculated as a safety feature; now the *Apollo 13* astronauts would have to execute this maneuver in order to direct their craft back to Earth.

In the LM things were already getting uncomfortable for Lovell, Haise, and Swigert. It didn't take long for the chilly air from *Odyssey* to drift into the LM, and the frigid temperature, mixed with the moisture from the crew's own breathing, was causing condensation on the LM's windows and instruments. As Lovell wiped the moisture off of the LM's windows, he saw from another perspective the scattered debris and the vaporous cloud swirling around *Aquarius*. Haise looked at Lovell and said, "Well, we're not going to be able to wipe that away, are we?"

Lovell was forced to assess their situation almost moment by moment, and he realized that *Odyssey*'s temperature was only going to drop lower. He suggested that Swigert gather as much food and water as he could from the CM before its temperature dropped below freezing. As Swigert pushed himself through the tunnel connecting the two vessels, he could see his own breath. He grabbed a bunch of food packets from the storage locker in the lower equipment bay and let them float freely while he filled drinking bags with water from the potable tank. But some of the water spilled on Swigert, soaking his feet. Resignedly, he figured that the way things were going, his feet would surely freeze before they would dry.

Lovell was less concerned about Swigert's wet feet than with the obstructed view from the LM windows. The swirling debris and gas outside the LM would make it difficult to navigate when he needed to prepare for a precise engine burn. To complete the burn he would need to determine the exact position of the craft by sighting specific stars through his alignment optical telescope (AOT).

Mission Control was concerned about it too. Even the slightest miscalculation of the ship's true attitude during the free-return trajectory burn could send *Aquarius* and *Odyssey* plummeting into the moon's surface. To get a clearer view, Lovell attempted to move the joined ships through the haze, but manually maneuvering the LM with the CM attached proved extremely difficult. When Lovell pushed the hand controller gently forward for thrust *Aquarius* pitched violently up and to the left. It didn't take Lovell and Haise long to realize that the LM's normal center of gravity was now greatly changed with the CM attached. The only way to get an accurate burn would be to figure out how to fly the combined vehicle.

Mission Control, meanwhile, was working on a host of other problems as well. While the Black Team continued to monitor current conditions, off-console retrofire officers (RETROs), flight dynamics officers (FIDOs), and guidance officers (GUIDOs) were at work calculating the exact short-burn maneuver that would get *Apollo 13* on the free-return trajectory. Next, they had to get approval for the PC + 2 burn. Glynn Lunney then asked Capcom Jack Lousma to brief the Apollo crew on what they needed to do.

First, Lousma told Lovell, the astronauts would have to execute a 16-foot-per-second burn. Then the crew would have to power down the LM to conserve oxygen and power. The all-important PC + 2 burn would follow 18 hours later.

Jim Lovell, inside the crowded Aquarius. The LM was not very comfortable when carrying the two astronauts it had been designed for; with all three crew members aboard, it was very cramped.

After the procedure was clearly outlined, Lousma and Lovell agreed to start the first burn in one hour.

But Lovell still hadn't been able to steer the vehicle out of the cloud of debris and gas. Lunney dispatched astronauts John Young and Ken Mattingly to the spacecraft simulators to see whether they could devise a maneuver that would help Lovell. Even backup LM pilot Charlie Duke, still at home sick with the measles, was called in, but to no avail. Finally, Lovell gave up and realized that they'd all just have to hope for the best when the coordinates for the burn were entered into the LM's navigation computer.

After running a series of tests in the LM simulator, Young and Duke, with the help of Grumman engineers in New York, determined that the craft's autopilot function could hold the joined crafts steady during the burn maneuver. It was time to start. Lovell was reminded to deploy the landing gear of the LM so that its spidery legs were not in the path of the engine. The burn was complete 37 seconds later.

In *Aquarius*, Lovell peered at the instrument panels while controllers on the ground anxiously reviewed the same data on their consoles. To the relief of all, the burn had gone perfectly. *Apollo 13* was safely on its slingshot course around the moon.

In the Mission Evaluation Room (MER) at Mission Control, dozens of technicians were trying to devise the best LM power-down scenario to keep the astronauts alive and well and get them back to Earth. No one questioned that the problem was a supreme challenge: they needed to figure out how to transform a craft that was designed to support two men for two days into one that would support three men for four days. They didn't have much time to come up with the right projections—as soon as the astronauts completed the short burn, they would begin powering down *Aquarius*.

The situation may have been dire, but Lead Flight Director Gene Kranz had remained optimistic about the Apollo crew's survival from the very beginning of the crisis. In Henry Cooper's *Thirteen: The Apollo Flight That Failed*, Kranz explains the reasons for his hopefulness. "I had the utmost confidence in the LM and in the flight controllers," Kranz said. "I knew that the life-support system was good, the communications were good, and the guidance system was good, and that it could make long rocket burns. I was sure it would prove to be a reliable spacecraft."

The LM, Kranz believed, would be an excellent "lifeboat" for the *Apollo 13* crew. The greatest concern was not over the craft's capabilities but over the dwindling levels of oxygen, water, and electricity available to the crew. Mission Evaluation Room director Don Arabian and his staff of 50 to 60 men prided themselves on knowing the most minute details of the millions of systems built into the CM and the LM. When something went wrong, it was their job to find out why. Thus, in the midst of the *Apollo 13* crisis, it was up to the MER team—and the Grumman experts—to determine exactly how the astronauts would need to parcel out their consumables to get home alive.

Since the LM operated on batteries instead of fuel cells, the oxygen supply would not be drained to produce power; the crew would have an ample supply of oxygen. Water, on the other hand, was extremely scarce. Not only did the astronauts need water to drink, but they also needed a substantial amount for the electronics cooling system. At the time of the explosion, *Aquarius* contained 338 pounds of water. During the first hour in the LM 6.3 pounds of water were consumed. Quick figuring indicated that at that rate the water supply would be depleted 23 hours sooner than the length of the fastest route home. The astronauts might have been able to survive a day without water; the equipment could not.

Power consumption was equally critical. Normally the LM used about 50 amperes of electricity per hour. Subtracting what had already been expended, the MER crew figured that the craft had just 2,000 ampere-hours of electricity left. That gave the crew only about 40 hours of power. This meant that to complete the burn maneuvers, the LM's normal minimal usage, 20 amperes per hour, had to be reduced to a scant 15. Arabian and his staff carefully reviewed every system and every component in the spacecraft to determine where they could save power. The questions they needed to ask were agonizing, and they needed

immediate answers. Could certain operating systems be turned off without adversely affecting the performance of the craft? Could electrical usage be lowered without compromising the effectiveness of the systems?

As daunting as the water and power problems were, an even greater dilemma arose: if the air in the LM was not filtered almost immediately, the astronauts would asphyxiate from the carbon dioxide buildup coming from their own breathing. *Aquarius* was equipped with two lithium hydroxide canisters, which filtered out the harmful carbon dioxide in the cabin air, but the number of hours the crew would occupy the LM exceeded the filters' capabilities. *Odyssey* had plenty of the canisters, but the filters from *Aquarius* and *Odyssey* were not interchangeable. Somehow, someone had to figure out how to use the CM canisters in the LM. The task fell to Mission

As the damaged spacecraft flew behind the moon, Haise and Swigert snapped photos while Lovell resigned himself to the fact that he would probably never step on the lunar surface. Fred Haise later commented, "We were . . . disappointed and sick that we could not land on the moon. We had lost a mission. We had worked a lot of years, and seemingly for nothing."

Control's Crew Systems Division. Armed only with parts available on the *Apollo 13* spacecraft, the team set to work solving the problem.

The scheduled PC + 2 burn was approaching, but Ground Control had not yet found a way for Lovell to make visual sightings from the LM. Finally, guidance officer Kenneth Russell suggested a simple solution: Lovell could check the guidance platform alignment against the position of the sun. Though it was hardly a pinpoint of light, like other stars in the sky, it would be enough to help Lovell position the spacecraft. It worked. Despite some maneuvering difficulties, Lovell and Haise finally lined up the sun in the crosshairs of the alignment optical telescope.

At 6:15 P.M. *Apollo 13* began its journey around the far side of the moon, where the ship would lose radio contact with the ground for about 25 minutes. Before the communications blackout, Gold Team Capcom Vance Brand radioed the data for the PC + 2 burn. At 76 hours, 42 minutes, and 7 seconds into the crippled mission, Lovell looked out the window and saw the sun set behind the moon. Two minutes later, the radio signal was lost.

About 20 minutes later Haise noticed a pocked gray orb outside the spacecraft—his first close view of the moon. He and Swigert stood transfixed at the LM's triangular windows, pointing out craters and hills that defined the moon's surface. Lovell, who had been here before and had seen the moon from an even closer perspective, relinquished his spot near the window to his crewmates.

It was a disheartening moment for Lovell; he knew that this was as close as he was ever going to get to stepping onto the moon. No one had made any promises to the *Apollo 13* crew about getting another shot at a moon mission. And Lovell knew that plenty of astronauts were waiting for their first opportunity to go into space. In the midst of his fourth mission, Lovell knew that his time to reach the moon had passed.

The commander of the *Apollo 13* mission watched the moon itself pass by as *Odyssey* and *Aquarius* circled behind it. And then he breathed a sigh of relief. After all, the endangered crew and their crippled spacecraft were now also on their way home.

When the astronauts jettisoned the service module, they could at last see the extent of the damage caused by the explosion. This photo was taken by Jack Swigert. The sight shook up the crew of Apollo 13; they began to wonder if the heat shield, which would protect the astronauts from the 5,000°F temperatures of reentry, had been damaged by the blast as well.

7

"*ODYSSEY,* HOUSTON STANDING BY, OVER"

"One of the things I deliberately excised from my mind was how cold it was going to get [in the spacecraft] and how tough it was going to be from a survival standpoint. When we were going to power down, we were going to power down to a level [where] these guys were going to live inside a freezer."
—Eugene Kranz, Lead Flight Director, *Apollo 13*

WHILE FRED HAISE and Jack Swigert were transfixed by the views graciously provided by *Aquarius*'s windows, Jim Lovell was preoccupied with the upcoming PC + 2 burn. A few minutes after *Apollo 13* emerged from behind the moon, Lovell hailed Mission Control to confirm that the radio signal was transmitting again, and then he began reviewing the procedures for the upcoming burn, now a little more than an hour away.

Haise and Swigert were still at the windows, clicking away and absorbing as much as they could of the view outside. Finally a slightly annoyed Lovell exclaimed, "Gentlemen, we have a PC + 2 burn maneuver coming up. Is it your intention to participate in it?" Disappointedly Haise answered, "This is our last chance to get these shots. We've come all the way out here—don't you think they're going to want us to bring back some pictures?" Lovell was sympathetic, but he also knew it was time to get back to business. "If we don't get home," Lovell pointed out, "you'll never get them developed." Haise and Swigert stowed their cameras and returned to their stations.

Over the next hour Capcom Vance Brand read off the LM power-up sequence that the ground crew had developed, and the flight crew began the process of flipping the appropriate switches. Before the countdown began, Brand informed the crew that their third-stage rocket, jettisoned early in the mission and deliberately aimed at the moon, had just made impact. The seismic instruments left on the moon by the *Apollo 12* crew were transmitting perfect readings to Ground Control. It was a small consolation; one aspect of the *Apollo 13* mission had gone as planned after all. Lovell glanced out the window, hoping to see moon dust kicked up by the rocket's impact. Instead he saw what would be his final view of a tiny mountain on the edge of the Sea of Tranquility—Mount Marilyn, the one he had named for his wife when he was a crew member aboard *Apollo 8.*

Haise's voice announcing "10 minutes to burn" nudged Lovell back to the task at hand. Finally, Brand gave Lovell the command: "Go for the burn." Lovell pushed an ignition button, and all three men felt *Aquarius* shudder. The procedure went off perfectly, and at the 4.5-minute mark Lovell shut down the engine and heaved a sigh of relief. When the free-return burn had been achieved six hours earlier, Lovell had been grateful that the spacecraft had been put on a course for Earth's atmosphere. The horrifying alternative would have been to orbit the moon endlessly, in view of everyone on the ground, as a ghastly "perpetual monument to the space program." Now that the PC + 2 burn had also been successful, the astronauts were on a trajectory for reentry—and a splashdown in the Pacific Ocean.

Lovell had hoped that the crew could immediately begin the LM power-down sequence and then get some sleep. But other chores needed attention. First, the astronauts had to perform a manual PTC (passive thermal control) maneuver, which would roll the ship so that one side wasn't always in the heat of the sun, a rotation similar to that of a chicken on a barbecue spit. Only after that could they begin the tedious LM power-down, and then the crew could rest. Haise took the first three-hour "sleep shift," drifting up to *Odyssey* while Lovell and Swigert stayed on watch in *Aquarius*.

In the cool, darkened LM a weary Lovell was caught up in his own thoughts. Right after the explosion he had told Haise and Swigert that the accident couldn't have come at a worse time in the mission, since the spacecraft was too far from Earth to turn around. Now Lovell realized that he had been wrong. Had it happened while they were in lunar orbit, *Aquarius*'s engines might not have been powerful enough to release the craft from the moon's gravitational hold. And had the SM exploded while he and Haise were on the moon, Swigert would have died while orbiting above them in *Odyssey*, and he and Haise would have been stranded on the lunar surface, awaiting the same fate. Lovell decided now that they had been extremely fortunate; in fact, if the explosion had to occur, it couldn't have come at a better time in the mission.

With all but the essential systems powered down in the LM, *Aquarius* seemed to grow colder each minute, and the astronauts were hardly dressed for a sudden temperature drop in their normally balmy spacecraft. Lovell guessed that the temperature in *Odyssey* was close to freezing as well.

Despite the chilly air and the constant noise in the LM,

Jim Lovell attempts to sleep in the chilly command module as Apollo 13 *heads back toward Earth.*

Lovell and Swigert managed to steal a few catnaps. Before long, Lovell heard Capcom replacement Jack Lousma telling him to rouse Haise from his slumber and change sleep shifts. When Lovell floated up to the CM to wake Haise, he noticed that the sleeping astronaut was surrounded by a thin layer of warm air. Lovell realized that without the presence of gravity, warm air would not rise the way it did in Earth's atmosphere. It seemed to be a good way to retain warmth—the catch was that you had to stay perfectly still. Once Haise stirred, his warm atmospheric blanket dissipated.

With Lovell and Swigert now in the "bedroom suite" of *Odyssey*, Haise enjoyed being alone in *Aquarius*. He described his view of the fading moon to Capcom Lousma, and then Lousma asked Haise for a carbon dioxide reading. Haise was slightly alarmed when he saw that the carbon dioxide gauge read 13. Under normal conditions the level would not rise above a 2 or 3, and any reading above 15 meant that the crew would begin to feel the effects of carbon dioxide poisoning.

On the ground, Ed Smylie, chief of the Crew Systems Division, had been working with his staff since the previous night (it was now late on Tuesday) on an improvised device that would allow the astronauts to fit the square-holed lithium canisters from *Odyssey* into the round-holed opening of the LM's environmental control system. Crew Systems not only had to devise a solution using only the materials on board the spacecraft, but they also had to find a way to describe the complicated task to the three astronauts, who would be working only with verbal directions and no illustration. As Joe Kerwin took over at Capcom, he earned the daunting job of talking the crew through the process.

Just before dawn on Wednesday morning Lovell drifted through the tunnel connecting *Odyssey* and *Aquarius*. Haise was surprised to see Lovell awake so soon, but Lovell complained that it was just too cold to sleep in the CM. Swigert joined his colleagues soon after. When Kerwin learned that all three men were awake, he figured that they might as well get started on the filter.

First Lovell, Haise, and Swigert were given a list of items to scavenge from the ships—odd items, including a space suit hose, duct tape, plastic bags, and the cardboard checklist covers from the astronauts' instruction manuals. After the items were assembled, and after more than an hour of instructions from the ground, the canister was ready to test. Swigert made the replacement, then pressed his ear against the open end of the canister. The slight hiss of air pulling through the filter vents confirmed that the contraption worked. In a matter of moments the carbon dioxide level in the spacecraft dropped. Yet another life-threatening crisis had been averted.

All that remained to do now was endure the cold, clammy atmosphere of the LM until the time came to move back into *Odyssey* for the flight home. Lovell and Swigert tried to get some sleep again. Then another problem arose: Capcom Brand advised Haise that the pressure in the LM's helium tank was rising. The helium, normally used to move the engine fuel into the combustion chamber for engine burns, could not exceed 1,800 pounds per square inch of pressure. If it did, the pressure-relieving burst disk would blow, and the helium would vent into space. Although the situation was not dire, it did mean that the LM's descent engine might not have enough fuel to execute another burn maneuver should one become necessary.

Brand also told Haise that the spacecraft's trajectory was "shallowing," or going off course. If it wasn't corrected, the ship would either skip off the atmosphere (like a stone across a pond) and enter a permanent orbit around the sun, or the opposite would happen: it would come in too steeply, and the extreme gravitational pressure of reentry would likely kill the astronauts. Even worse, no one at Mission Control had yet figured out why the trajectory was shallowing. In any event, another midcourse burn would be necessary later in the evening. Brand anticipated that the helium burst disk would blow at about 105 hours into the mission, and he scheduled the burn for 104 hours. Both men hoped that even if the helium disk blew earlier, they could still pull off the burn.

Once again, as he had been in the early hours of the morning, Haise was the solitary skipper of *Aquarius*. He took out an audiotape of the popular song "Aquarius" and blared it loud enough for the ground team to hear it through the air-to-ground loop. He wandered over to the LM's window, hoping to get a good view of the receding moon, when suddenly he heard a "thump-bang-shudder" similar to the one they'd heard during the explosion. Haise knew that the sound couldn't be the helium tank just yet, and he eliminated the possibility that it had come from the SM.

At the same time, TELMU Bob Heselmeyer noticed that the LM's battery two level had started to drop quickly—just like the oxygen tank had in *Odyssey* on Monday night. If battery two was dying, then only three-quarters of the LM's already carefully rationed power remained. As it turned out, the battery had exploded. But even counting the sharp reduction of power, this latest problem was not life threatening.

Lovell awoke at about 3 P.M., and Haise and Brand quickly brought the commander up to date. By late afternoon all three astronauts were awake and preparing for the next engine burn. Brand radioed the LM power-up sequence. To conserve power, the computer and mission timer would remain off; Lovell performed the burn manually and once again demonstrated his stellar navigational skills.

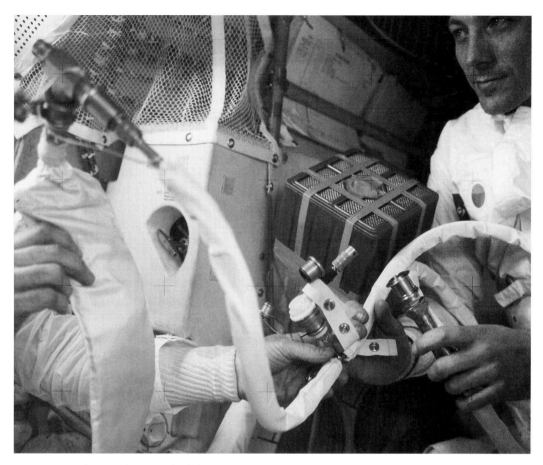

Now 36 hours from splashdown, Lovell, Haise, and Swigert settled back in the LM. The next major task was the all-important CM power-up. For almost three days Gene Kranz's White Team had been sequestered in room 210 of Mission Control, working out the procedures to get the CM up and running using only the two hours of power its three reentry batteries would supply. Finally the finished checklist—dozens of pages long and entailing hundreds of steps—was ready. Kranz told his team to go home and get some sleep, something none of them had done since Monday night.

Haise, meanwhile, had begun to feel feverish. All three astronauts had carefully rationed their water consumption

Fred Haise helps to put together the makeshift air filter needed to remove carbon dioxide from the LM's atmosphere. At Mission Control, members of the ground crew had been working for days to construct the filter from items that were available on the damaged spaceship.

and were becoming dehydrated. Mission doctors had warned them that in the weightless atmosphere of space this condition might create a toxin buildup that could lead to kidney infections. Haise had all the symptoms of such an infection, including fever, achiness, pale skin, and excruciating pain when he tried to urinate. Lovell noticed how poorly Haise looked and asked him if he was okay.

Just then, the helium disk blew. Capcom Lousma acknowledged the readout with Lovell, but they could do nothing to fix it, so Lousma suggested that the crew relax. Haise and Swigert fell asleep; Lovell continued to discuss the remaining maneuvers with Lousma.

Things didn't stay quiet for long. The ship's trajectory was shallowing again. On the ground a host of engineers and controllers gathered in a workroom to discuss how the *Apollo 13* crew could safely jettison the damaged service module and then the LM before they positioned *Odyssey* for reentry. The safest way to jettison the SM, they determined, was to have Lovell and Haise remain in *Aquarius* while Swigert returned to *Odyssey*. Just before separation Lovell would fire *Aquarius*'s thrusters to move the whole vehicle forward. Then Swigert would release the service module, and Lovell would fire the LM's thrusters again in reverse, moving *Odyssey* and *Aquarius* away from the released SM. The LM jettison would happen in almost the same way, but instead of initiating another power-draining thrust, the crew would leave the passageway between the two crafts partially pressurized, so that *Aquarius* would "pop" away upon release.

On Thursday evening—the crew's final night in space—Lovell visited the cold, lifeless *Odyssey* and immediately noticed the frosty buildup of moisture on its windows, walls, and instrument panels. Did this mean that the wires, bulbs, and other electrical devices behind the instrument panels were also soaked? When it came time to power up *Odyssey* and this moisture began to flow freely, it could cause a short in the system, which might shut

everything down again. And the CM was the only craft that could get the crew home. Lovell hoped there would be no surprises in store.

The long wait for the power-up checklist did little to alleviate Lovell's anxiety. Splashdown was 18 hours away, and he knew that it would take hours for Ground Control to read the procedures while Swigert copied each one down and then read it back to Control to confirm accuracy. After several requests by Lovell, Brand finally radioed that the list was ready. Reading it to Swigert took nearly two hours, from 7:30 to 9:15 P.M. In less than 12 hours the crew would begin to power up *Odyssey*. If all went well, Lovell, Haise, and Swigert would splash down three hours after that.

The astronauts tried to get some rest, but *Odyssey* was

Apollo 13's lunar module, jettisoned shortly before the astronauts reentered the Earth's atmosphere. "Farewell, Aquarius, and we thank you," one Ground Control member said as the ship drifted away.

just too cold, and the men were too anxious and weary to sleep soundly. Eventually, Deke Slayton asked Flight Director Milt Windler whether the crew could begin bringing the lunar module back online earlier than originally planned, since power consumption had been so carefully controlled. Windler checked with TELMU Jack Knight and confirmed a "go." Slayton got on the loop with Lovell. "Okay, Skipper," he said. "We figured out a way for you to keep warm. We decided to start powering up the LM now." Lovell and Haise got *Aquarius* running in less than half an hour. Almost immediately the LM began to warm up. The crew of *Apollo 13* was in the homestretch.

Shortly after sunrise on Friday morning the astronauts were back to work. The first job was to jettison the service module. Lovell hit the LM's thrusters to move the ships forward, and Swigert pressed the button that released the SM. Having completed the maneuver, all three men scrambled to windows, cameras in hand, for their first glimpse of the service module as it floated away from them.

What Lovell saw left him momentarily speechless. As the SM cylinder rolled, he could see that an entire panel was missing, from the base of the module to its engine. Compartment four had been ripped open, and he could see its exposed innards—pieces of insulation, dangling wires, and a charred hole where oxygen tank two should have been. As Lovell tried to describe the view to Mission Control, Swigert snapped photo after photo.

Once the astronauts saw how badly damaged the service module was, they began to worry that the command module's heat shield, which had been near the SM, had also been damaged. No one expressed these concerns aloud, however. A damaged heat shield was an extremely grave problem, but there was nothing anyone could do to fix it. There was no point in dwelling on it.

About three hours after the SM was released, Capcom Joe Kerwin told Swigert to power up *Odyssey*, a process that took another half hour. With less than two hours until splashdown, Lovell and Haise prepared to abandon *Aquarius*. They grabbed a few souvenirs, joined Swigert in *Odyssey*, and sealed the hatch to the tunnel. Swigert flipped the switch to jettison *Aquarius*. As the LM floated away from the command module, Kerwin spoke softly, "Farewell, *Aquarius*, and we thank you."

All that remained of the *Apollo 13* spaceship was its tiny command module. The last chore for the crew before they settled in for reentry was to check the capsule's attitude. All was well. *Odyssey* blazed toward Earth at about 25,000 miles per hour. Just before the radio blackout that occurred during reentry, Jack Swigert spoke to the ground crews. "I know all of us here want to thank all you guys for the very fine job you did," he told them. Then communications went silent. Mission Control could do nothing now but wait until radio contact was restored. The friction generated from the spacecraft plummeting through the upper layer of Earth's atmosphere would push temperatures past 5,000°F across the capsule's heat shield. If the shield was intact, the crew would be fine. No one wanted to think about the alternative: the crew, having survived everything else, would burn up during reentry.

After four minutes of radio silence, Gene Kranz directed Kerwin to hail the flight crew. "*Odyssey*," said Kerwin, "Houston standing by, over." Fifteen seconds elapsed with no reply. Kerwin tried again. Another 15 seconds passed; Mission Control heard only static. Another 30 seconds passed, and controllers began shifting nervously in their chairs as they stared at their console screens. At last, after more than five minutes had passed, Jack Swigert's voice came through. "Okay, Joe," he said.

Mission Control was eerily silent for a few seconds. Then the room erupted in loud cheers and applause. Kranz pumped his fist into the air. Kerwin took a deep breath and acknowledged Swigert's message. "Okay, we read you, Jack," he said with relief.

Soon after, *Odyssey*'s parachutes opened, and on the big

As the crew of Apollo 13 *arrives safely on the aircraft carrier* Iwo Jima, *the flight directors—Gene Kranz, Glynn Lunney, and Gerald Griffin— puff on cigars in celebration of the spacecraft's successful return. The* Apollo 13 *mission ended at 12:07 P.M. April 17, 1970.*

screen in Mission Control an image of the capsule floating toward the ocean appeared. The scene was broadcast on televisions around the world. Despite all the serious trajectory problems, *Odyssey* splashed down at 12:07 P.M. just three miles from the recovery ship—the most accurate landing in the Apollo program's history up to that time. Now, bobbing in the Pacific Ocean, Lovell turned to his crewmates and said simply, "Fellows, we're home."

Almost immediately, rescue crews from the aircraft carrier *Iwo Jima* lifted the astronauts via helicopter to the ship. Military frogmen involved in the rescue later described the blast of frigid air that escaped from the capsule when they opened the hatch.

Once safely aboard *Iwo Jima*, the astronauts received a

congratulatory call from President Richard Nixon. The next call the president made was to Marilyn Lovell. He asked whether she would like to accompany him to Hawaii to meet the carrier transporting her husband and his crewmates to land. With tears of joy and relief, Marilyn Lovell agreed.

Lovell, Swigert, and Haise are debriefed by Wernher von Braun (right) and Deke Slayton, the director of flight crew operations. Immediately after the safe return of the Apollo 13 *astronauts, NASA began investigating the accident to prevent it from happening again.*

8

A SUCCESSFUL FAILURE

"I agreed with the solution. If it worked, we could launch on time. If it didn't, we would probably have to replace the tank, and the launch date would slip. None of the launchpad test crew knew that the wrong thermostat was in the tank, or thought what would happen if the heaters stayed on for too long."

—James A. Lovell, testifying at the hearing investigating the cause of the *Apollo 13* explosion

WITHIN HOURS OF *Apollo 13*'s triumphant splashdown, NASA administrator Tom Paine appointed a committee to investigate the cause of the accident. The 15-member panel, which included Edgar Cortright, director of NASA's Langley Research Center, and *Apollo 11* commander Neil Armstrong, toiled for two months before it presented its findings. The photographs taken by the flight crew had ruled out some of the possible causes: for example, the destruction was too "clean" to have been caused by a meteor or other foreign

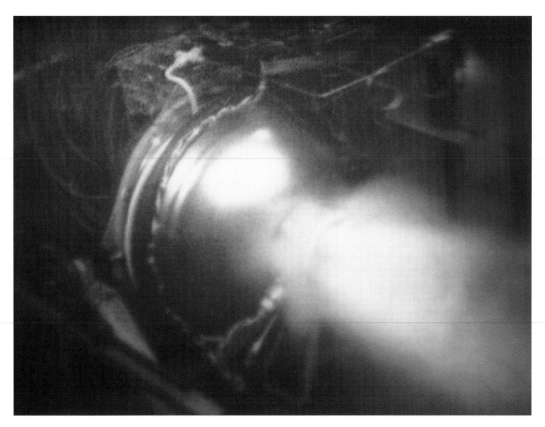

This photo was taken during a NASA test to determine the effect of fire inside a service module oxygen tank. The lid of the tank is blowing off, allowing the gas to escape.

object colliding with the SM. The damage seemed to have been caused by an internal malfunction.

Most of the panel members also concluded that the accident was not the result of one catastrophic failure, but rather of a series of failures dating to 1965. They discovered that a communication lapse that year prevented engineers from correctly modifying the cryogenic tank's thermostatic heater to accept a voltage increase, from 28 to 65 volts DC. The oversight alone would not have caused the disaster, but it was certainly a contributing factor.

Next, the panel reviewed the history of oxygen tank two. Originally intended to be part of *Apollo 10*'s SM in 1968, it had been removed so a newer, advanced-design tank could be installed. During removal the tank had dropped, but an external inspection revealed no damage, so the tank was upgraded and installed in *Apollo 13*'s service module.

Then, three weeks before *Apollo 13*'s scheduled launch, the ship and crew underwent a routine countdown demonstration test (CDT), a dress rehearsal of the actual ignition and liftoff. As part of this drill the cryogenic tanks were fully pressurized and the astronauts donned their flight suits. The hours-long test went smoothly—until it was time to drain the cryogenic tanks to complete the CDT. The two hydrogen tanks and oxygen tank one drained without incident, but the second oxygen tank did not. At the time, however, everyone assumed that since the venting tube used to drain the tank was not the same one that supplied oxygen to the fuel cells and environmental systems, the glitch would have no effect on the tank's operation during the actual mission.

After consulting with others, including mission commander Lovell, the engineers decided simply to "boil off" the oxygen in tank two by using the tank's internal heater

The below photo, taken after a test of the Teflon-insulated wiring used inside the service module's oxygen tanks, shows how the insulation cracked at high temperatures.

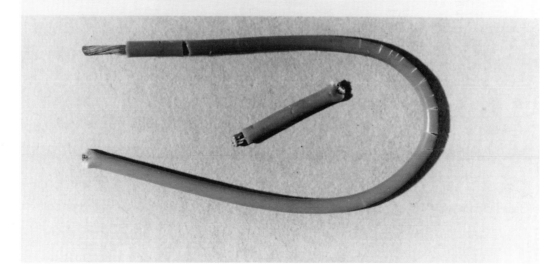

752°F FOR 60 MINUTES

and fan. In *Apollo: The Race to the Moon*, authors Charles Murray and Catherine Bly Cox explain the reasoning behind the decision: "This was considered the best procedure because it reproduced the way the system would work during flight: heating the liquid oxygen, raising its pressure, converting it to gas, and expelling it through the valves and pipes into the fuel cells where, in flight, it would react with the hydrogen."

A safety switch inside the tank was supposed to cut off the heaters should the temperature rise above a safe 85°F. Even so, a technician was assigned to monitor the eight-hour process. Sure enough, the gauge showed that the temperature had reached 85°F. What was unknown, however, was that this temperature was the gauge's highest limit. The technician, therefore, had no way of knowing that the temperature in the tank had actually risen to a scalding 1,000°F. This happened because the safety switch, which triggered correctly, was almost instantly welded shut by the 65-volt power surge it was not designed to handle. As a result, the shutoff did not occur, and the heaters continued to run, baking the Teflon insulation on the wiring inside the tank until it cracked open, exposing the wires.

Seventeen days later, when *Apollo 13* was in orbit and Sy Liebergot requested a cryo stir, those exposed wires inside oxygen tank two shorted and a spark ignited the remaining Teflon insulation. The ensuing fire and pressure buildup blew off the neck of the tank, turning 300 pounds of pure oxygen into gas and flooding the number four compartment of the service module. In seconds the highly flammable gas exploded, ripping off the craft's external panel. The raging fire that resulted was quenched in the vacuum of space. Had it not been for that vacuum, the entire spacecraft—and the crew with it—would have exploded in flames.

As for the shallowing problem with the LM, the investigators determined that extra thrust was created by the

heated vapors emitted from *Aquarius*'s cooling system. The force of the emission was just enough, over the 240,000 mile journey back to Earth, to alter the trajectory of *Apollo 13*.

NASA's Apollo program resumed only after engineers and designers had upgraded the cryo-tank thermostat voltage switches, improved the wire coating, and added an extra oxygen tank in a separate area of the service module. Lovell himself was in Mission Control on January 31, 1971, as *Apollo 14*—the mission he was originally scheduled to command—lifted off for its journey to the Fra Mauro highlands of the moon.

What stunned Jim Lovell, Fred Haise, and Jack Swigert on Friday, April 17, after their splashdown and recovery was the overwhelming support they had garnered during their ordeal in space. The entire world, it seemed, had been praying and hoping for their safe

Aboard the Iwo Jima, *Jim Lovell reads about* Apollo 13 *in a Honolulu newspaper. The crew of the damaged spaceship were amazed that their flight around the moon and back had captured the attention of the world.*

return. The *Christian Science Monitor* summed up the reaction: "Never in recorded history has a journey of such peril been watched and waited out by almost the entire human race." America's relief—and that of the world— was echoed in newspaper headlines across the globe, like the one splashed across the front page of the *New York Times*: "Astronauts Land Gently on Target, Unharmed by Their Four-Day Ordeal."

Medical checkups showed that the astronauts were in relatively good condition despite Haise's infection and their dehydration and physical exhaustion. On their first evening back on Earth, Lovell and Swigert enjoyed a welcome dinner of shrimp, prime rib, and lobster with the officers of the *Iwo Jima* (Haise was confined to the ship's sick bay). The next day President Nixon made a brief stop in Houston, where he presented the nation's highest award, the Medal of Freedom, to the flight directors of Mission Control. Then he continued to Honolulu, where Marilyn Lovell, Mary Haise, and Dr. and Mrs. Leonard Swigert (Jack's parents) were reunited with their exhausted but safe husbands and son. The president conferred the Medal of Freedom on the *Apollo 13* crew as well, and he praised the three astronauts and the Mission Control crew for having transformed great adversity into a resounding triumph.

Jim Lovell never did set foot on the moon. After he removed himself from NASA's lunar flight rotation, he transferred to the agency's new space shuttle program, where he worked with McDonnell Aircraft to design the ship's extensive instrument panel. He later retired from NASA and formed his own company, Lovell Communications, based in Lake Forest, Illinois.

Neither Fred Haise nor Jack Swigert ever reached the moon either. Haise became a test pilot for the space shuttle *Enterprise* before leaving NASA to become president of Grumman Technical Services. Both Lovell and Haise also traveled the lecture circuit with other colleagues, including former Lead Flight Director Gene Kranz.

After serving on the Science and Aeronautics Committee of the House of Representatives, Jack Swigert resigned from NASA in 1978 and decided to enter politics. He lost a race for the Senate, but on November 2, 1982, he was elected a Colorado congressman. Sadly, the victory was short-lived: before he could be sworn in, Swigert was diagnosed with a rare form of cancer. He died on December 27, 1982.

More than 30 years have passed since Jim Lovell, Fred Haise, and Jack Swigert concluded their perilous journey on *Apollo 13*. Though all three agreed that their mission had failed, Lovell has since referred to it as a "successful failure," in which hard but necessary lessons were learned. But when the command module splashed into the ocean on April 17, 1970, no one was thinking about failure. Instead, they focused on the way in which the efforts, ingenuity, and talent of a highly dedicated ground and flight crew had averted what would have been one of the worst disasters in the history of space exploration.

The key had been a refusal to believe that the three astronauts would not return to Earth. The elation shared by *Apollo 13* ground and flight crews with countless people around the world was eloquently summed up by NASA's Tom Paine, who convened the investigation following the accident. "There has never been a happier moment in the United States space program," Paine declared. "Although the *Apollo 13* mission must be recorded as a failure, there has never been a more prideful moment."

CHRONOLOGY

April 11, 1970

1:13 P.M. CST   *Apollo 13* lifts off from Cape Canaveral

3:48 P.M.   The Saturn V third-stage rocket propels *Odyssey* and *Aquarius* out of Earth's orbit toward the moon

5:14 P.M.   *Odyssey* completes the transposition maneuver, docking with *Aquarius*

April 12, 1970

7:53 P.M.   Jack Swigert fires *Odyssey*'s engines to leave the free-return trajectory

April 13, 1970

8:56 P.M.   The crew of *Apollo 13* completes a televised broadcast; White Team EECOM Sy Liebergot requests a cryo stir of the craft's hydrogen and oxygen tanks

9:07 P.M.   Inside the Service Module, oxygen tank two explodes. Swigert radios to Mission Control: "Houston, we've had a problem."

10:50 P.M.   The crippled *Odyssey* forces mission commander Jim Lovell and Fred Haise to power up *Aquarius*. Meanwhile, Swigert executes a complete power-down of *Odyssey*.

April 14, 1970

2:43 A.M.   Lovell fires *Aquarius*'s engine to set a free-return trajectory for Earth

6:15 P.M.   Radio contact between Mission Control and *Apollo 13* is lost during the spacecraft's orbit behind the moon. Haise and Swigert get their first close-up look at the moon, now just 136 miles away

8:40 P.M.   *Aquarius*'s engine fires to execute the PC + 2 burn

April 15, 1970

3:38 A.M.   The flight crew begins assembling adapted canisters to eliminate carbon dioxide buildup in *Aquarius*

2:23 P.M.   Battery two in *Aquarius* explodes; the spacecraft loses one-quarter of its remaining power

10:31 P.M.   *Aquarius*'s engine is fired again to correct a course drift

April 17, 1970

7:14 A.M.   After the service module is jettisoned, the *Apollo 13* astronauts are able to view the extensive damage to the vessel

10:43 A.M.   The *Apollo 13* crew returns to *Odyssey* and jettisons *Aquarius*

11:53 A.M.   *Odyssey* begins reentry into Earth's atmosphere

12:07 P.M.   *Odyssey* splashes down in the Pacific Ocean; recovery crews transport Lovell, Haise, and Swigert to the waiting aircraft carrier *Iwo Jima*

APPENDIX

AN INTERVIEW WITH GENE KRANTZ, LEAD FLIGHT DIRECTOR FOR THE *APOLLO 13* MISSION

AT FIRST GLANCE, the crew cut ex-fighter and former test pilot Eugene Kranz projects an image of being in full control, more driven to solving problems than being alarmed by them. In his years at Mission Control in the Johnson Space Center in Houston, Texas, he was just that kind of person. But Kranz is also a man who relies on a strong faith and who values the importance of family (he has six children and seven grandchildren). He believes in freely expressing his emotions, and he has always trusted his own instincts to solve problems.

Kranz flew combat missions during the Korean War, and joined NASA just as the space program was getting under way. In a career that spanned more than 30 years, Kranz was almost a permanent fixture in Mission Control, acting as flight director and division chief for Flight Control from Project Mercury until the end of Project Apollo. As director of Mission Operations, beginning in 1983, Kranz was also responsible for "all aspects of mission design, testing, planning, training, and spaceflight operations." He is perhaps best remembered for leading his flight team—nicknamed the "Tiger Team"—during the *Apollo 13* crisis, and for bringing the crew safely back to Earth. Kranz spoke with the author of this book about his time with NASA, and especially about the *Apollo 13* mission, one of the defining moments of his career.

■ **Did you ever consider becoming an astronaut?**

■ **GENE KRANZ:** I saw opportunities in Mercury [Mission] Control, and decided that was the direction I was gonna go. . . . I wasn't thinking a long-term career, I was thinking of pure fun. I could work a bunch of missions a year and these guys [the astronauts] . . . would be tied to only one, so [by working in Mission Control] I could see more of the big picture.

■ **In 1960–61, wasn't there some trepidation about putting a man atop a rocket that was actually a reconfigured missile and firing him into the air?**

■ **GK:** No. . . . When I was at Holliman [Air Force Base] there was a literal explosion in the business of flight testing. You'd put people in balloons and take them up to

100,000 feet and they'd bail out in a parachute and they'd free fall for 95,000 feet before opening up the parachute. You had the Zero Launch program, where you'd strap people on an airplane and fire them off the back of a truck. . . . People were doing crazy things all the time, and this was basically the environment I grew up in. . . . so putting a person on top of a rocket was just the next step.

■ **You described working in the space program as "pure fun," but you look like a very intense man. Can you explain that in a little more detail?**

■ **GK:** Well, it was [intense], but . . . each mission was a challenge all in itself, and it was just a question of getting smart enough to rise to the next challenge. . . . So it was a series of escalating steps that we were taking within the program, each one more difficult, more challenging . . . and that was about as exciting as it could get for some kid in the '60s.

■ **What did the *Apollo 1* tragedy do to you, to the space program?**

■ **GK:** There wasn't any question that we were really smacked in the face by the *Apollo 1* fire, but again, it was the kind of thing that I believe set the stage for [making] the mental determination to make sure that something like that never happened again. . . .

[On the] Monday following the accident . . . we accepted the accountability for [it] because I firmly believe that there was something we should have caught, something we missed. We had too much "go-itis," and [after the accident I held a meeting where] we established two words in the vocabulary of Mission Control: "tough" and "competent." "Tough" meaning we will never again shirk from any responsibilities we have and will never stop learning, [and "competent" meaning that] from now on, the Mission Control team will be perfect. And when [I] finished [giving] that talk to my troops in there, we all went home. I wrote [those words] on the blackboard, and . . . [they] were never erased until the final Apollo mission [had ended].

■ **When the *Apollo 11* astronauts landed on the moon July 20, 1969, what was that moment like for you and your team working that mission?**

■ **GK:** None of us in the team at that instant had enough time even to reflect on what we did, because the instant we [landed the astronauts] we had less than 120 seconds to decide [whether] it was safe to remain on the surface [of the moon]. . . . We had [to make] a series of what they call "stay-no stay" decisions . . . so we were in constant motion from . . . the time we started the descent through a very difficult landing. We finally got down with less than 17 seconds of fuel remaining. While the rest of the world is rejoicing, we're busting our butts to go through the stay-no stay decisions and make sure we're making all the right [choices]. . . .

We had a lot of tense moments. . . .We had problems in communications, we had problems in navigation, we had a small electrical problem, we had to work around the computer program alarms, and then we just about ran out of gas. . . . It was only after[wards], . . . while I was walking over to the press conference, [that] I had the ability to just reflect for the first time that, by God, we landed today.

■ **When the problem occurred with *Apollo 13*, how far into the situation were you before you realized how serious it was?**

■ **GK:** The thing that was frustrating to me, and this is both during [and] after the fact, was that it took so long to recognize the magnitude of the problem. We were totally confident in the design of the spacecraft. We had redundant systems, . . . so no one would ever believe that we were going to lose our cryogenic oxygen, both tanks, all three fuel cells—a major portion of the components in the service module. So it took . . . almost 20 minutes for me to come to the point where I realized that we had moved from a lunar landing problem . . . to a survival problem, and what irritated me was the length of time it took to come to that realization.

■ **What did you think was the greatest threat to the crew's safe return?**

■ **GK:** Doubt. . . . The press described me as the "arrogant optimist" because I would never concede the loss of this crew. I mean, it was completely unfathomable that anyone would ever believe that we're gonna end up with this crew stranded somewhere [in space]. So it was a question of absolutely convincing everyone who had to work a problem, no matter how miniscule, that (a) they were going to [get the] right answers that they needed, and (b) they were going to add all these answers together, and we were going to get this crew home. So my concern was that somewhere, someone along the line would turn around the bit I had set in everyone's mind that this crew's coming home.

■ **How did you deal with the possibility that, even though** *Apollo 13* **had returned safely to Earth, the astronauts might burn up in the atmosphere if the heat shield had been damaged?**

■ **GK:** It's one of the things I never worried about. It's sort of like [saying] I fly airplanes and I wear a parachute. I never worried, "Is that parachute going to open up if I have to use it?" It's just really a question of saying [that] if it doesn't, that's the breaks of the game. . . . Hey, there's some point where you've done the best—I call it the "Hum" or the Human Factor—where the human beings have done everything that they conceivably can, we've given it our best; and then you have to be ready to live with whatever happens. And we were [at that point] during reentry.

■ **What was it like during the capsule's reentry into Earth's atmosphere, when because of the radio blackout Mission Control lost contact with the astronauts?**

■ **GK:** The blackout was—is—the toughest time, I think, for every controller, because you know now that there's no more givebacks, the crew's on their own; you just pray that everything you gave the crew and everything you did was right, and that all these pieces added up together. And you just hope that all the procedures you gave them were sufficiently clear so they can execute them with no more help.

And then [with *Apollo 13*] we got that extended blackout period, and I think everyone who . . . worked that mission will never forget it. It was probably the most difficult time of the entire mission. When the parachutes finally did open, I think . . . the emotional climax for all the people that were involved there was so great [that] I found myself crying. It was really tough to keep going. A lot of the controllers in the room just stood up and it was absolute silence, and then you had a brief cheer, and then everybody sat down because we still had the landing to get through.

■ **It seems so unfair that you couldn't even celebrate at the time.**

■ **GK:** That was the beauty of writing my book [*Failure Is Not an Option: Mission Control from Mercury to Apollo 13 and Beyond*], because literally I had all the people [who worked on the space project] together, and we'd put a tub of beer out . . . and we'd talk. And we had more stories, some of which I'd never heard before. So it was a very fun time, and it was a good opportunity to get back together and really let it soak in 30 and 40 years later what [we had done].

■ **All these many years later, even after all this time, when you think about** *Apollo 13*, **what does it do to you?**

■ **GK:** I think there's a very unique feeling amongst the people [who worked that mission]—I call them "the brotherhood" in my book. You know, a lot of people look back on their high school years . . . and the kinds of things they did. I think the beauty of this thing here is [that] every time I look back at it, there isn't anything I would have done differently, and I think the respect and true affection that the people felt for each other . . . was unique. I've only seen it in the military; I've never seen it in the civilian side of the business except among the flight controllers. I believe that's the relationship [among] the controllers and between the controllers and the crews.

I speak a lot with Fred Haise. . . . and I speak with Jim Lovell, speak with Ken Mattingly, T.K.—we have get-togethers, and I think the one thing that's constant there is the respect we all have for each other. And that's really what you go into life hoping you're going to get, and it's the one thing you have left when you go out.

BIBLIOGRAPHY

Apollo 13: "Houston, We've Got A Problem." Produced by Finley-Holiday Film Corporation. 60 min. Finley-Holiday Films Space & Science Series. Videocassette.

Apollo 13: To The Edge and Back. Produced and directed by Noel Buckner and Rob Whittlesey for WGBH Boston, TV Asahi Japan, and Central Independent Television U.K. 87 min. WGBH Educational Foundation, 1994. Videocassette.

Cooper, Henry S. F., Jr. *Thirteen: The Apollo Flight That Failed.* Baltimore: Johns Hopkins University Press, 1972.

Crouch, Tom D. *The National Aeronautics and Space Administration.* New York: Chelsea House, 1990.

Doomed Mission: The True Story of Apollo 13. Produced by Simitar Entertainment, Inc. 90 min. 1995. Videocassette.

Hacker, Barton C., and Charles C. Alexander. *On the Shoulders of Titans: A History of Project Gemini.* NASA History Series, no. 4203 (1977).

Kennedy, Gregory P. *Apollo to the Moon.* New York: Chelsea House, 1992.

Kranz, Gene. *Failure Is Not an Option: Mission Control from Mercury to Apollo 13 and Beyond.* New York: Simon and Schuster, 2000.

Lovell, Jim, and Jeffrey Kluger. *Lost Moon: The Perilous Voyage of Apollo 13.* New York: Houghton Mifflin, 1994.

Mellberg, William F. *Moon Missions: Mankind's First Voyages to Another World.* Plymouth, Mich.: Plymouth Press, 1997.

Pellegrino, Charles R. and Joshua Stoff. *Chariots for Apollo.* New York: Avon Books, 1985.

Reeves, Robert. *The Superpower Space Race: An Explosive Rivalry through the Solar System.* New York: Plenum Publishing, 1994.

Shepard, Alan, and Deke Slayton. *Moon Shot: The Inside Story of America's Race to the Moon.* Atlanta: Turner Publishing, 1994.

Swenson, Loyd S. Jr., James M. Grimwood, and Charles C. Alexander. *This New Ocean: A History of Project Mercury.* NASA History Series, no 4201 (1989).

Voyt, Gregory. *Apollo and the Moon Landing.* Brookfield, Conn.: Millbrook Press, 1991.

INDEX

JUDY L. HASDAY, a native of Philadelphia, Pennsylvania, received her B.A. in communications and her Ed. M. in instructional technologies from Temple University. A multimedia professional, she has had her photographs published in many magazines and books. As a successful freelance author, Ms. Hasday has written several books, including an award-winning biography of James Earl Jones and biographies of Madeleine Albright and Tina Turner. She is the coauthor of *Marijuana*, a book for adolescents that presents the facts about and dangers of using the drug. Recent works include a book about the Japanese attack on Pearl Harbor.

JAMES SCOTT BRADY serves on the board of trustees with the Center to Prevent Handgun Violence and is the vice chairman of the Brain Injury Foundation. Mr. Brady served as assistant to the president and White House press secretary under President Ronald Reagan. He was severely injured in an assassination attempt on the president, but remained the White House press secretary until the end of the administration. Since leaving the White House, Mr. Brady has lobbied for stronger gun laws. In November 1993, President Bill Clinton signed the Brady Bill, a national law requiring a waiting period on handgun purchases and a background check on buyers.

PICTURE CREDITS